Peterson's egghead's Guide to Geometry

Cara Cantarella

About Peterson's

Peterson's provides the accurate, dependable, high-quality education content and guidance you need to succeed. No matter where you are on your academic or professional path, you can rely on Peterson's print and digital publications for the most up-to-date education exploration data, expert test-prep tools, and top-notch career success resources—everything you need to achieve your goals.

For more information, contact Peterson's, 800-338-3282 Ext. 54229; or find us online at www.petersonsbooks.com.

ISBN-13: 978-0-7689-3662-9

Printed in the United States

10 9 8 7 6 5 4 3 2 1   15 14 13

Petersonspublishing.com/publishingupdates

Check out our Web site at www.petersonspublishing.com/publishingupdates to see if there is any new information regarding the test and any revisions or corrections to the content of this book. We've made sure the information in this book is accurate and up-to-date; however, the test format or content may have changed since the time of publication.

Table of Contents

Table of Contents

Table of Contents

Table of Contents

Table of Contents

Table of Contents

Before You Begin

Welcome to *egghead's Guide to Geometry*! My name is egghead, and I'll be your guide throughout the book.

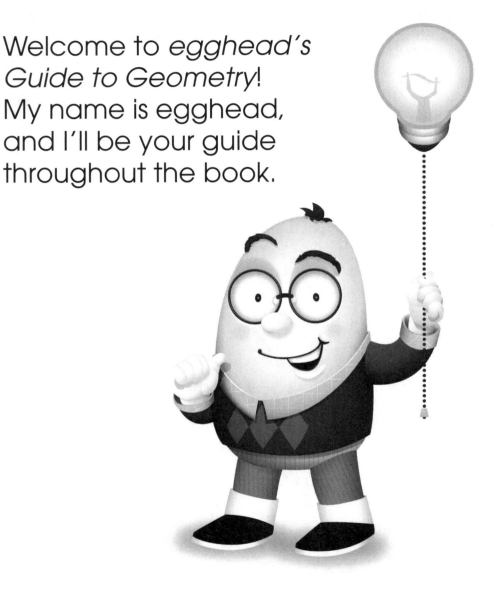

Before You Begin

This egghead's Guide was designed to help you learn geometry in a fun and easy way. Sometimes learning can be . . . well, boring. It can also be confusing at times. If it wasn't, we'd all have straight A's, right?

As your guide through the adventure of education, I'm here to make things a bit more enjoyable. I studied the boring books so you don't have to. I got straight A's and lived to tell about it. I understand this stuff, and you can too. In this guide, I'll show you what you need to learn to get to the next level.

Wherever I can, I explain things in pictures and stories. I break concepts down and teach them step by step. I try to stick with words that you know. I give examples from real life that you can relate to.

I want you to succeed, and I know you can do it!

In this book, we'll work together to improve your geometry skills and build your confidence. Confidence is very important, and it comes from trust. You can trust me as your guide, and most important, you can trust yourself. If your geometry knowledge isn't strong enough, let's do something about it!

How this book is organized

This book contains twelve chapters. We recommend you read them in order.

The introduction comes first.

Chapter 1 explains lines and points. These are basic parts of most geometry shapes.

Chapter 2 explains angles. Angles are important on their own, but they show up in other shapes, too.

Chapters 3, 4, 5, 6, and 7 focus on three main shapes. These are: quadrilaterals, triangles, and circles. We learn the properties of these shapes and how to find their perimeter and area.

Chapter 8 focuses on polygons. Here we look at figures with five or more sides.

Chapter 9 addresses irregular and multiple figures. We learn how to work with unusual shapes and answer questions about them.

Chapters 10, 11, and 12 describe three solid figures: cubes, rectangular solids, and cylinders. We learn their properties and how to find volume and surface area.

As you read each chapter, you'll find practice exercises along the way. Complete the exercises to practice what you've learned. The more you use your skills, the better they will "stick" in your mind.

Practice makes perfect!

To learn more

Ready for more practice? After you've finished this book, visit the egghead website at www.petersonspublishing.com. Click the egghead link for even more practice exercises. This book will get you off to a great start. The website can give you that extra geometry boost!

Find us on Facebook

You can find us on Facebook® at www.facebook.com/petersonspublishing. Peterson's resources are available to help you do your best in school and on important tests in your future.

Peterson's books

Along with producing the egghead's Guides, Peterson's publishes many types of books. These can help you prepare for tests, choose a college, and plan your career. They can even help you obtain financial aid. Look for Peterson's books at your school guidance office, local library or bookstore, or at www.petersonsbooks.com. Many Peterson's eBooks are also online!

We welcome any comments or suggestions you may have about this book. Your feedback will help us make educational dreams possible for you—and others like you.

Now that you know what's ahead, let's get started!

Introduction

Geometry is one of the main math subjects taught in American high schools today, along with pre-algebra, algebra I, algebra II, trigonometry, and calculus. Many colleges require knowledge of geometry in order to be admitted to the school. You may not ever need to take geometry classes in college, but for competitive schools, you will be required to pass standardized tests that have geometry questions in the math sections. In addition, understanding geometry may be a requirement for many career paths you might want to pursue.

Currently geometry questions appear on all of the following standardized tests:

• GED	• GRE
• SAT	• GMAT
• ACT	• MCAT

But you know all the official reasons for learning geometry. This book will help you realize the fun and practical use for this often-lamented math subject. Geometry is a fascinating topic that is involved in nearly all we do. Have you ever tried to decorate your room? How did you find out if the new television you bought would fit? How did you know your new bed would move easily through your bedroom door frame? Believe it or not, you used some form of geometry to find out.

Without an understanding of measurements, you might buy a dresser that's too big and blocks out the light from the window. Or, your couch might extend to the middle of the doorway. Oops!

You can also use geometry to figure out everyday problems. Many of these problems have to do with building, such as how to mitre a corner on a chest of drawers. Never plan to build a thing in your life? Well, geometry can come in handy for knowing whether a package will pass postal regulations and how to keep your mailing costs low. You might need it for calculating how much water to put in an aquarium. It can also be useful when purchasing carpet or renting an office that charges by the square foot. If you'd like to be able to check the charges and be sure you're getting what you pay for, a little geometry knowledge can be a big help.

In these next chapters, we focus on the main concepts of geometry that appear on many standardized tests. Each chapter has review questions that help you see how what you've learned might be applied in the real world. We start with the basics and build up to more difficult material as we go. While this book doesn't cover every possible topic—geometric proofs are excluded, for instance—it does go over the concepts you need to know to build a solid foundation in the subject. Who knows . . . it may even help you one day to build your own home!

Let's begin! Meet egghead!

Part 1

Plane Geometry

Chapter 1

Lines and Points

Hi! I'm egghead. I will teach the following concepts in this chapter:

What is a line?
What is a point?
Other geometry terms—
   dimensions, planes,
   postulates, and theorems
Line segments
All about length

Finding lengths
What about units?
Midpoints and rays
Parallel lines
Intersecting lines
Bisectors
Transversals

What is a line?

Many of the figures that you'll see in geometry are made up of lines.

In geometry, the word **line** always refers to a straight line.

Technically speaking, lines go on forever. They extend into space both ways.

In geometry, this is the symbol for a line:

The arrows mean that the line goes on forever.

Naming lines

To name a line, put a letter near it. The line in this figure is line *a*.

a

What is a point?

A **point** is a specific location on a line.

We almost never see points by themselves in plane geometry. We usually see them on lines.

Examples

Here is a point on line *p*.

p

Here is a point on line *z*.

z

Naming points

In geometry, when we see points on a line, they usually have names.

Examples

This line has points *B* and *C*.

This line has points *D*, *E*, and *F*.

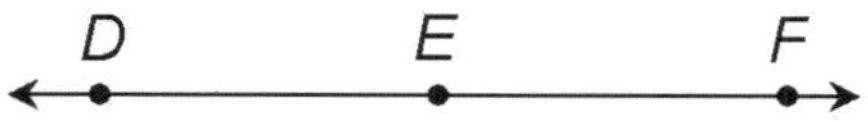

Points in geometry are usually named with letters, not numbers. Most often, points are in capital letters.

The points are used to show *locations* on the line.

Practice Questions

1. Name the lines shown below. Name the first line *x*, the second line *y*, and the third line *z*.

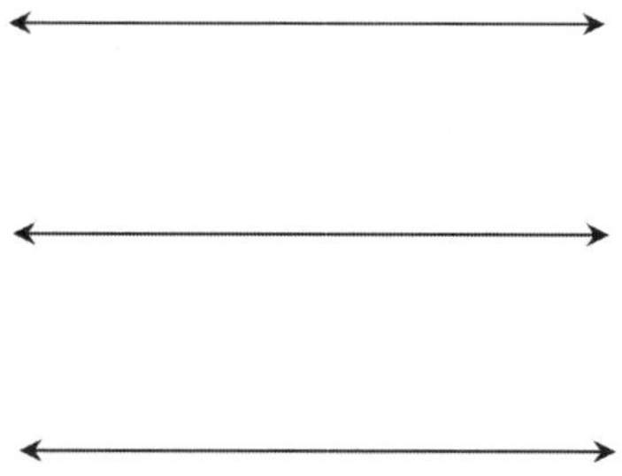

2. Name the lines shown below. Name the first line *q*, the second line *r*, and the third line *s*.

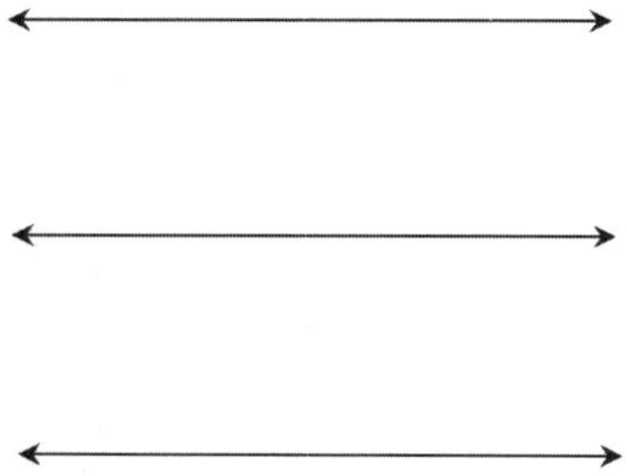

3. Draw in points *P* and *Q* on the line below.

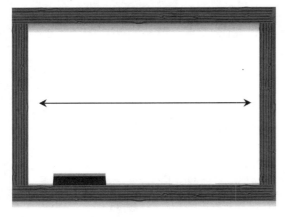

4. Draw in points *R* and *S* on the line below. Name the line *m*.

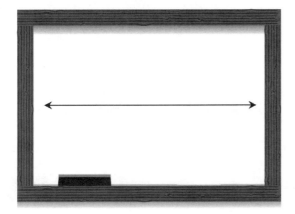

 **Solutions**

1.

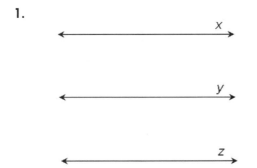

egghead's Guide to Geometry

2.

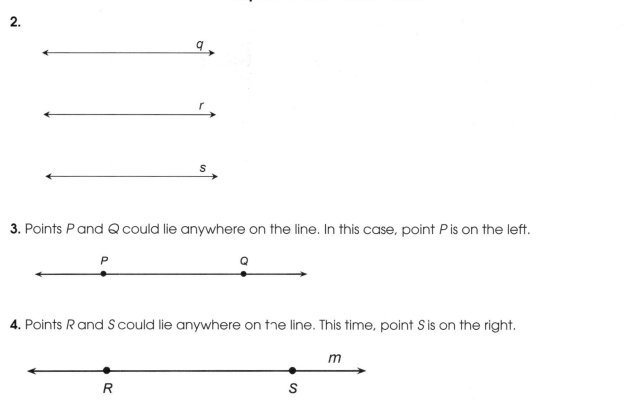

3. Points *P* and *Q* could lie anywhere on the line. In this case, point *P* is on the left.

4. Points *R* and *S* could lie anywhere on the line. This time, point *S* is on the right.

Other geometry terms: dimensions, planes, postulates, and theorems

Lines and points are two of the basic building blocks in geometry. We will talk a lot more about them in the rest of this chapter. Before we do, there are a few other concepts you should know.

Dimensions

To measure items in geometry, we refer to different dimensions, such as length, height, and width. Points are units that have no dimension. They have no size or length; they just indicate locations on a line.

Lines in geometry have one dimension. Many common shapes, such as squares and other flat figures, have two dimensions. There are also solid figures, which have three dimensions. We'll discuss those in Part 2.

Planes

A plane is a special component of geometry that makes up a flat surface. A plane is a set of three or more points that are not on the same line. Just as lines extend infinitely both ways, planes extend infinitely in all directions. Here is a picture of how planes are usually drawn:

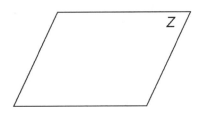

Note the capital letter *Z* in the upper right corner. This is the label for the plane, plane *Z*.

Postulates

An example of a postulate is that two points determine a line.

Geometry does not just consist of shapes and figures. It also operates by certain rules. These rules define what we know about geometry and how we calculate certain measurements. Geometry postulates are statements that are accepted as true. They do not require proof, and they cannot be proven. They are taken as givens.

Theorems

A theorem is like a geometry postulate, except that it can be proven. Whereas postulates cannot be proven and are assumed to be true, theorems can be shown to be true through a series of logical steps.

One common theorem that we will discuss in a later chapter is the Pythagorean theorem. It explains the relationship between the three sides of a right triangle. Using the theorem, you can find the length of a missing side of the triangle.

Line segments

Portions of lines are sometimes called **line segments**. Line segment means "a part of a line."

In this figure, the line segment starts at point *B* and ends at point *C*.

In this figure, the line segment starts at point *D* and ends at point *E*.

The figure below contains three line segments.

One line segment starts at point *D* and ends at point *E*.

Another line segment starts at point *E* and ends at point *F*.

The third line segment starts at point *D* and ends at point *F*.

Symbols

In geometry, there is a special symbol that means line segment.

The line over *BC* means line segment.

Line segment *BC* is written as $\overline{BC}$.

Practice Questions

1. What is the name of the line segment below? Use the symbol for line segment.

2. What is the name of the line segment below? Use the symbol for line segment.

3. Name the line segments shown in the figure below.

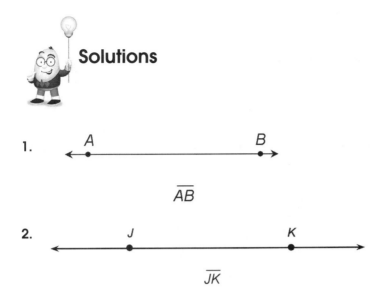

Solutions

1. $\overline{AB}$

2. $\overline{JK}$

3. The line segments are $\overline{LM}$, $\overline{MN}$, and $\overline{LN}$.

All about length

To show the length of a line segment, we write in the measurement.

Examples

The length of $\overline{BC}$ is 4.

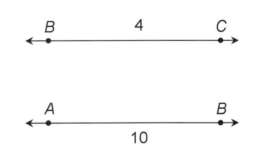

The length of $\overline{AB}$ is 10.

Finding lengths

Sometimes, lengths are not marked. We can use what we know to find the missing lengths.

Examples

In the figure below, we know that the length of $\overline{AC}$ is 5 and the length of $\overline{CB}$ is 5.

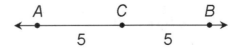

We can use what we know to find the length of $\overline{AB}$. If $\overline{AC}$ is 5 and $\overline{CB}$ is 5, then $\overline{AC} + \overline{CB} = 10$.

egghead's Guide to Geometry

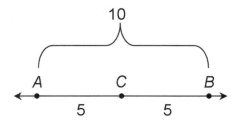

We add $\overline{AC}$ plus $\overline{CB}$ to find the length cf $\overline{AB}$.

We can also use subtraction to find missing lengths.

The figure shows that the length of $\overline{GJ}$ is 13 and the length of $\overline{HJ}$ is 9.

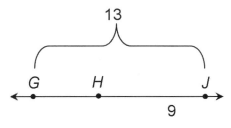

To find the length of $\overline{GH}$, subtract the length of $\overline{HJ}$ from the length of $\overline{GJ}$: 13 – 9 = 4.

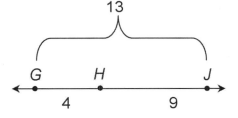

The length of $\overline{GH}$ is 4. A short way to write measurement is using the letter m.

In this case, $m\overline{GH} = 4$.

Practice Questions

1. What is the length of line segment *CD* below?

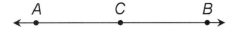

2. The length of $\overline{AC}$ is 6. The length of $\overline{CB}$ is 6, too. Write in the lengths on the figure below.

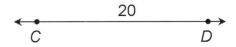

3. Write in the length of line segment *QS* in the figure below.

4. Write in the length of line segment *BD* in the figure below.

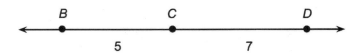

5. Write in the length of line segment *YZ* in the figure below.

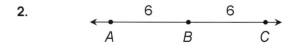

Solutions

1. The length of $\overline{CD}$ is 20.

2.

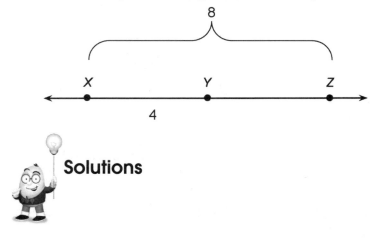

3. We add $\overline{QR}$ plus $\overline{RS}$ to find the length of $\overline{QS}$.

4. We add $\overline{BC}$ plus $\overline{CD}$ to find the length of $\overline{BD}$.

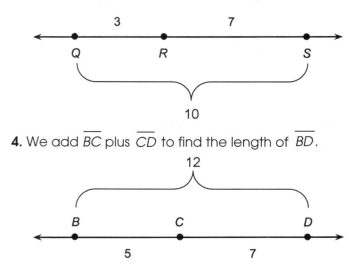

5. The correct answer is shown below.

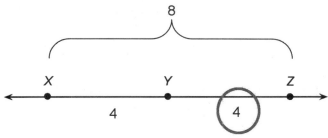

We subtract $\overline{XY}$ from $\overline{XZ}$ to find the length of $\overline{YZ}$.

$$\overline{YZ} = \overline{XZ} - \overline{XY}$$
$$\overline{YZ} = 8 - 4$$
$$\overline{YZ} = 4$$

Excellent work!

What about units?

Normally, when we measure lengths, we use some sort of unit. Lengths might be in feet or inches, for example. When we use units to measure length, we simply write the unit given. For instance, the measure of $\overline{MN}$ is 6 feet.

6 feet

Sometimes we must multiply the lengths of the sides of a figure. In this case, when we multiply feet by feet, the answer must be given as square feet: 6 feet × 4 feet = 24 square feet. A short way to write this is 24 ft^2.

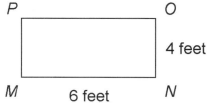

With solid figures, we might multiply three measurements. In this case, the units are expressed as cubic units. If we were to multiply 6 feet × 4 feet × 2 feet, the answer would be 48 cubic feet. We could write this in shorthand as 48 ft^3.

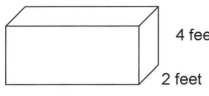

4 feet

2 feet

6 feet

We will work more with 2-dimensional and 3-dimensional shapes in later chapters.

Midpoints and rays

Before we leave this discussion of points and length, there is one more type of point you need to know about. This is called the **midpoint.**

Examples

Midpoints are points that fall halfway between two points on a line. Point *Y* is the midpoint of $\overline{XZ}$:

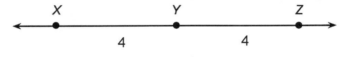

The length of $\overline{XZ}$ is 8. Point *Y* falls in the middle of $\overline{XZ}$.

Point *C* is the midpoint of $\overline{AB}$.

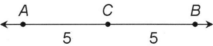

The length of $\overline{AB}$ is 10. Point *C* lies halfway between *A* and *B*.

Rays

A ray is section of a line that begins at one point but continues on forever. Here's an example:

This is ray *ST*. It could also be written this way:

$$\overrightarrow{ST}$$

Practice Questions

1. Circle the midpoint of the line below.

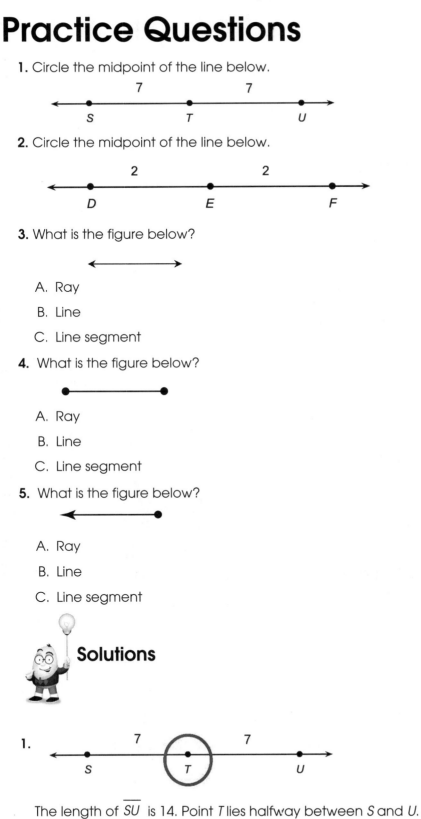

2. Circle the midpoint of the line below.

3. What is the figure below?

    A. Ray

    B. Line

    C. Line segment

4. What is the figure below?

    A. Ray

    B. Line

    C. Line segment

5. What is the figure below?

    A. Ray

    B. Line

    C. Line segment

Solutions

1.

The length of $\overline{SU}$ is 14. Point T lies halfway between S and U.

2.

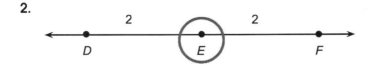

The length of $\overline{DF}$ is 4. Point *E* lies halfway between point *D* and point *F*.

3. The correct answer is B. The figure is a line.

4. The correct answer is C. The figure is a line segment.

5. The correct answer is A. The figure is a ray.

Parallel lines

In geometry, there are some special types of lines. The first type is called **parallel.**

Parallel lines are lines that don't cross each other. They never meet.

Examples

Some parallel lines look like this:

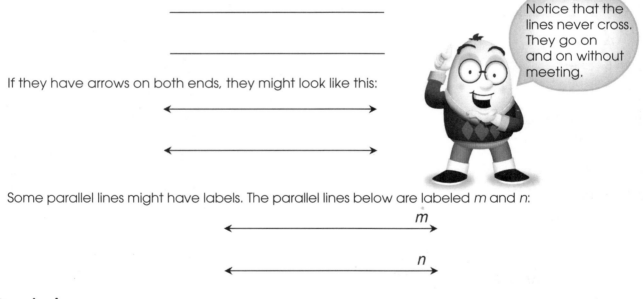

If they have arrows on both ends, they might look like this:

> Notice that the lines never cross. They go on and on without meeting.

Some parallel lines might have labels. The parallel lines below are labeled *m* and *n*:

Symbols

In geometry, almost every item has a symbol. The symbol for parallel lines is: $\|$

The parallel lines *m* and *n* in the previous example are shown with the symbol like this:

$$m \parallel n$$

This means that line *m* is parallel to line *n*.

Sometimes, the parallel line symbol is slanted, like this: //

Either type of symbol is okay to use. You'll probably see the ‖ symbol most often.

Intersecting lines

Along with parallel lines, in geometry there are also **intersecting lines.** Intersecting lines are lines that cross each other.

Examples

Here are some examples of what intersecting lines look like:

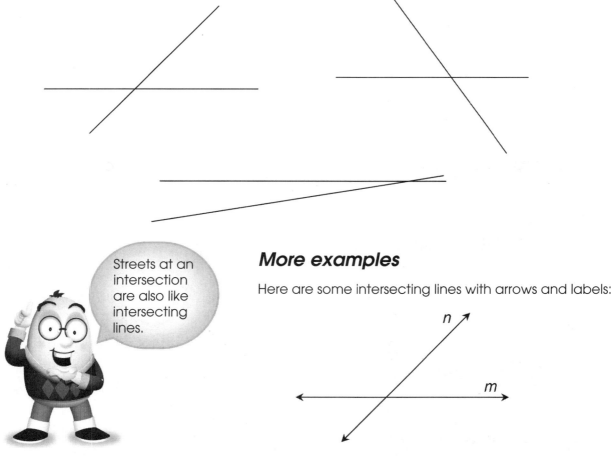

Streets at an intersection are also like intersecting lines.

More examples

Here are some intersecting lines with arrows and labels:

n

m

There aren't any specific symbols to show intersecting lines. In geometry, intersecting lines are among the very few items that *don't* have a symbol.

Practice Questions

1. Draw the symbol for "line *a* is parallel to line *b*."

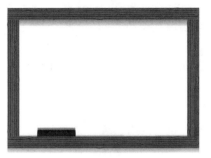

2. Draw line *a* and line *b* so that they are parallel.

3. Draw two lines, *j* and *k*, showing that *j* ‖ *k*. Also show that both lines go on forever.

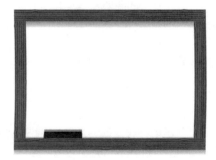

4. Draw lines *x* and *y* that intersect.

egghead's Guide to Geometry

5. Draw two lines showing that line *p* intersects line *q*.

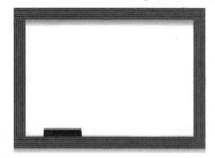

 **Solutions**

1. The symbol is: $a \parallel b$

2.

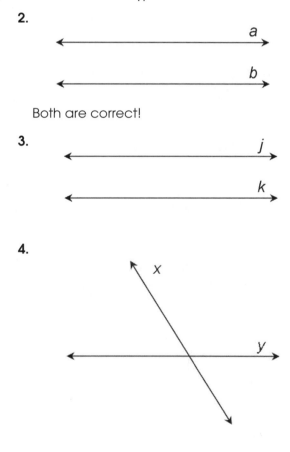

Both are correct!

3.

4.

5.

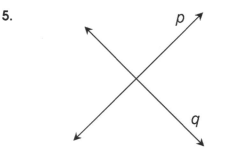

Bisectors

A special type of intersecting line is called a **bisector.** Bisectors, or bisecting lines, are lines that cross a line segment at its midpoint. Remember midpoints? Midpoints are points that fall halfway between two points on a line. Bisectors divide a line segment into two equal parts.

Examples

In the figure shown, line *m* is a bisector. It divides line segment *XZ* into two equal parts.

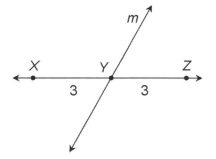

In this figure, $\overline{OP}$ bisects $\overline{EG}$. As the figure shows, $m\overline{EF} = m\overline{FG}$.

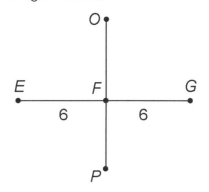

Practice Questions

1. Line segment VW bisects $\overline{RT}$ at point S. If the measure of $\overline{RT}$ is 8, what is the measure of segment RS?

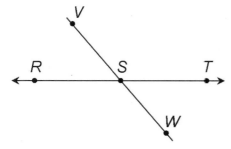

2. If $\overline{EG}$ bisects $\overline{OP}$, what is the measure of $\overline{OP}$?

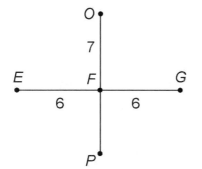

3. If line segment CD bisects line segment AB at point E, what is the measure of line segment EB?

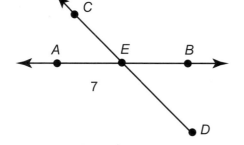

4. Using the same figure as above, determine the measure of AB.

5. In the figure shown, line segments HI and JK are the same length. Line segment HI bisects $\overline{JK}$ at point L. What is the length of $\overline{HI}$?

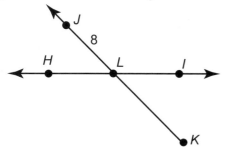

Solutions

1. Line segment VW bisects $\overline{RT}$ at point S. Therefore, $\overline{RS}$ and $\overline{ST}$ are equal. We are told that the measure of $\overline{RT}$ is 8. Divide $\overline{RT}$ into two equal parts: $8 \div 2 = 4$.

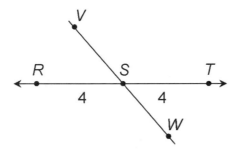

2. We are told that $\overline{EG}$ bisects $\overline{OP}$. The figure shows that $m\overline{OF} = 7$. Therefore, the measure of $\overline{FP}$ must also equal 7. Add together $m\overline{OF} + m\overline{FP}$ to determine the length of $\overline{OP}$: $7 + 7 = 14$.

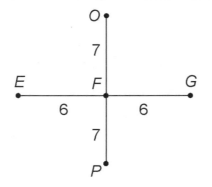

3. Line segment CD bisects line segment AB at E. That means E is the midpoint of line segment AB, which means line segments AE and EB are equal. We are told that the measure of line segment AE is equal to 7. So the measure of line segment EB is also 7.

4. We know that both line segments AE and EB measure 7. Adding them together gives us $7 + 7 = 14$. The measure of line segment AB is 14.

5. We are told that $\overline{HI}$ bisects $\overline{JK}$. The figure shows that line segment JL equals 8, which means $m\overline{JK} = 16$ $(8 + 8)$. We're told both line segments HI and JK are equal, so $m\overline{HI}$ is 16 also.

Transversals

Before we leave the subject of lines, there's one more type of line you should know about. It's called a **transversal.**

A transversal is a line that intersects two other lines.

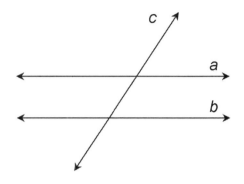

Technically, a transversal is a line that intersects two or more other lines in the same plane at different points.

What's important is that you recognize a transversal when you see one.

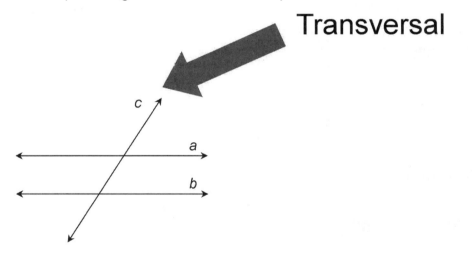

In this case, line *c* is a transversal. It crosses parallel lines *a* and *b*.

In geometry, we usually see transversals crossing two lines that are parallel.

Practice Questions

1. Draw a transversal, *x*, that intersects lines *y* and *z*. Show lines *y* and *z* as parallel.

2. Draw two parallel lines, *m* and *n*, crossed by a transversal, *t*.

3. One more time. Draw two lines, *p* ∥ *q*, crossed by a transversal, *s*.

4. Draw three parallel lines, *t*, *u*, and *v*, crossed by a transversal, *w*.

5. If a line crosses three lines that are not parallel, is it considered a transversal?

Solutions

1.

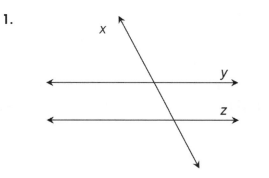

Line *x* could also slant the other way:

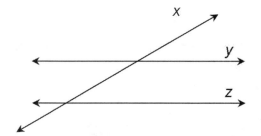

Many different options are correct.

2.

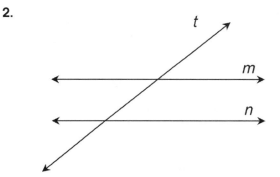

3. In this case, we have the symbol *p* ∥ *q*. This means line *p* is parallel to line *q*. Line *s* is the transversal. The lines might look like this:

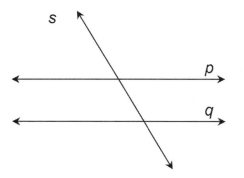

4. In this case, we have three lines *t*, *u*, and *v* that are parallel. So your drawing may look like this:

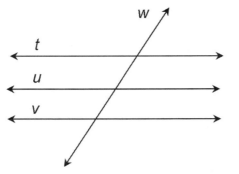

5. You were asked if a line that crosses three other lines that are not parallel is considered a transversal. The answer is yes, but only if all three lines are in the same plane, and the three lines are crossed at different points.

Chapter Review

1. Draw in points *X* and *Y* on the line below.

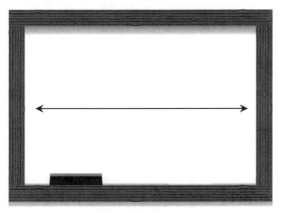

2. Name the line segments shown in the figure below.

3. Write in the length of line segment *JL* in the figure below.

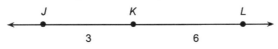

4. Write in the length of line segment *QR* in the figure below.

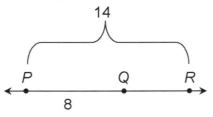

Chapter 1: Lines and Points

5. Circle the midpoint of the line below.

A   B   C

4   4

6. Draw two lines, *d* and *e*, showing that *d* ‖ *e*.

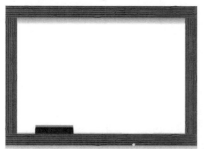

7. Draw lines *f* and *g* that intersect.

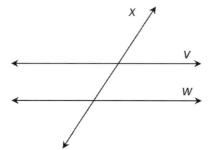

8. In the figure shown, which line is the transversal?

X

V

W

9. Draw two parallel lines, *b* and *c*, crossed by a transversal, *o*.

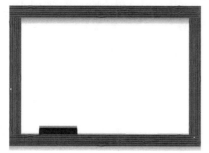

10. Draw two lines, $s \parallel t$, crossed by a transversal, r.

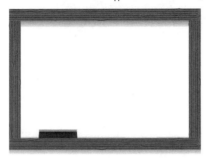

11. Amanda lives on the same street as Harvey and Carole. Their houses lie on a straight line. Harvey's front door is exactly 20 yards from Amanda's front door. Carole's front door is exactly 30 yards from Harvey's front door. Draw a figure that shows the distances between the three houses.

12. Amanda starts at her house and walks to Harvey's house. She stops at the door briefly to talk to Harvey, and then she continues to Carole's house. How many yards does Amanda walk?

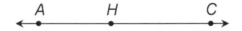

13. Scarlet's school is on the same street as the library. The two buildings lie on a straight line exactly 2 miles apart. After school, Scarlet rides her bike to the library to study. When she gets to the library, she realizes she forgot her backpack in her locker, so she rides back to school, picks up her backpack, and returns to the library. How many miles does Scarlet travel in these three trips?

14. A pharmacy, a diner, and a bakery are located on the same street, shown in the figure below. The diner is on the corner of a cross street exactly 10 miles from the pharmacy. The bakery is located 20 miles from the pharmacy. What is the distance from the bakery to the diner?

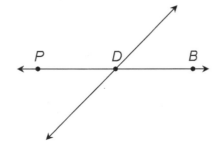

15. A wooden bar, *w*, bisects a steel beam, *s*, as shown in the figure below. If the steel beam measures 15 meters from end to end, what is the distance from one end of the steel beam to the wooden bar?

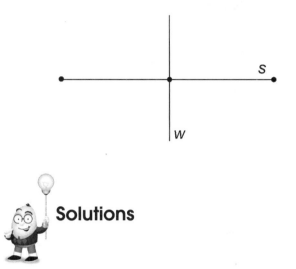

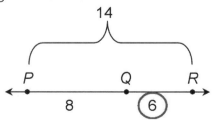

Solutions

1. Points *X* and *Y* could lie anywhere on the line.

2. The line segments are $\overline{FG}, \overline{GH},$ and $\overline{FH}$.

3. We add $\overline{JK}$ plus $\overline{KL}$ to find the length of $\overline{JL}$: 3 + 6 = 9. Line segment *JL* measures 9 units.

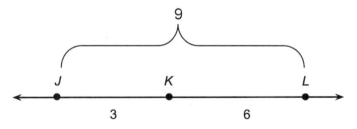

4. We subtract $\overline{PQ}$ from $\overline{PR}$ to find the length of $\overline{QR}$. In this case, 14 − 8 = 6.
Line segment *QR* measures 6 units.

5. The length of $\overline{AC}$ is 8. Point *B* lies halfway between *A* and *C*.

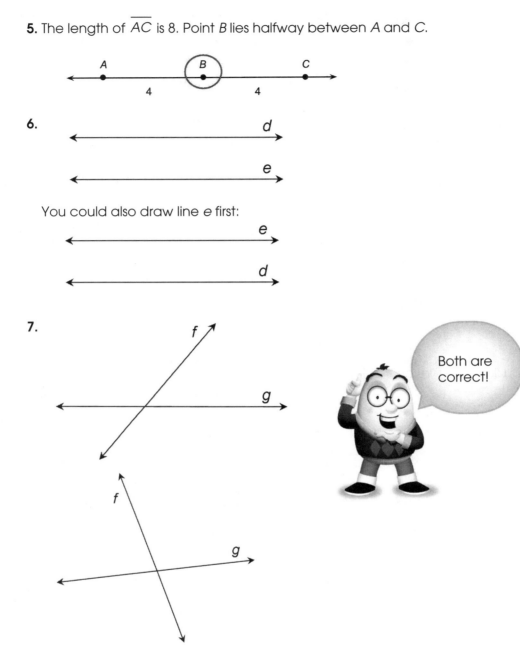

6.

You could also draw line *e* first:

7.

Both are correct!

8. In this figure, line *x* is the transversal.

Transversal

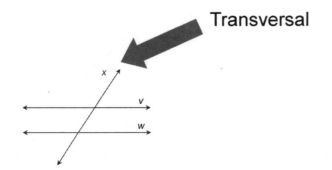

9. The lines might look like this:

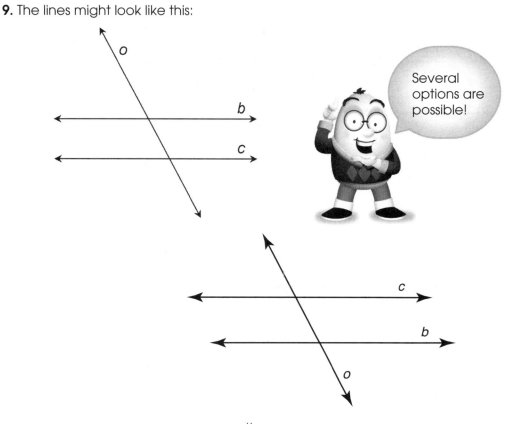

10. In this case, we have the symbol $s \parallel t$. This means line s is parallel to line t. Line r is the transversal. The lines might look like this:

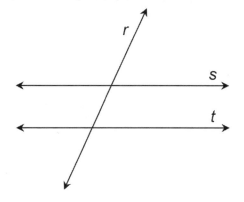

11. In the diagram, Amanda's house is shown by the letter A. Harvey's house is shown as H, and Carole's house is labeled C.

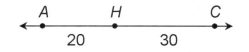

The length of line segment AH is 20; there are 20 yards between Amanda's house and Harvey's. The length of line segment HC is 30, as there are 30 yards between Harvey's and Carole's houses.

12. The correct answer is 50 yards.

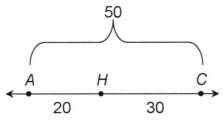

Amanda starts at her house, stops at Harvey's, and then continues to Carole's. So, she walks 20 yards + 30 yards, or 50 yards total.

13. The correct answer is 6 miles.

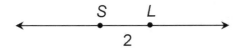

The figure shows the distance from Scarlet's school, *S*, to the library, *L*. Scarlet rides once from school to the library, then from the library to school, and a second time from school back to the library. Each time she makes the trip, she rides for two miles. The arrows in the figure show Scarlet's three trips:

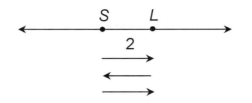

Scarlet travels 2 + 2 + 2 miles, or 6 miles total.

14. Use subtraction to determine the distance from the bakery to the diner. If the bakery is 20 miles from the pharmacy, and the diner is 10 miles from the pharmacy, this means the diner lies halfway between the bakery and the pharmacy. The diner is 20 − 10 = 10 miles from the bakery.

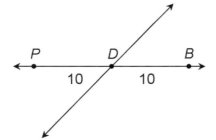

15. The correct answer is 7.5 meters.

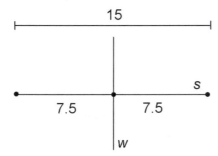

The steel beam measures 15 meters total. The wooden bar bisects the steel beam, crossing the steel beam exactly in the middle. The wooden bar lies halfway between the ends of the steel beam, or 7.5 meters from each end.

Chapter 2

Angles

Hi! I'm egghead. I will teach the following concepts in this chapter:

What is an angle?
Parts of angles
Naming angles
Degrees
Adding angle measures
Bisecting rays
Degrees in a line
Supplementary angles

Right angles
Complementary angles
Perpendicular lines
Acute and obtuse angles
Adjacent angles
Vertical angles
Congruence
Corresponding angles

What is an angle?

Angles are formed when two lines intersect.

Examples

Remember the intersecting lines we saw in the last chapter?

These lines form angles *a, b, c,* and *d:*

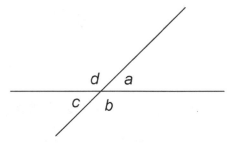

These lines form angles *w, x, y,* and *z:*

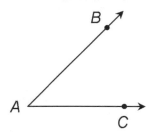

Where lines cross, they form angles.

Parts of angles

In the last chapter we learned that a ray is a line that starts at an endpoint and goes on forever in one direction.

An angle is made from two rays that have the same beginning point. This angle has ray *AB* and ray *AC.*

egghead's Guide to Geometry

This angle has ray *DE* and ray *DF*.

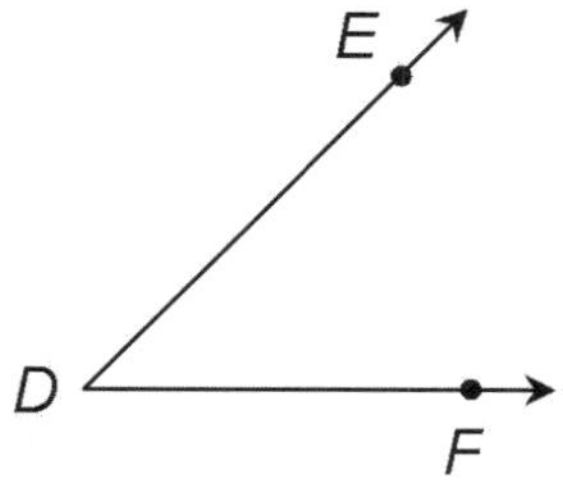

The point of the angle is called the **vertex.**

The angle below has vertex *A:*

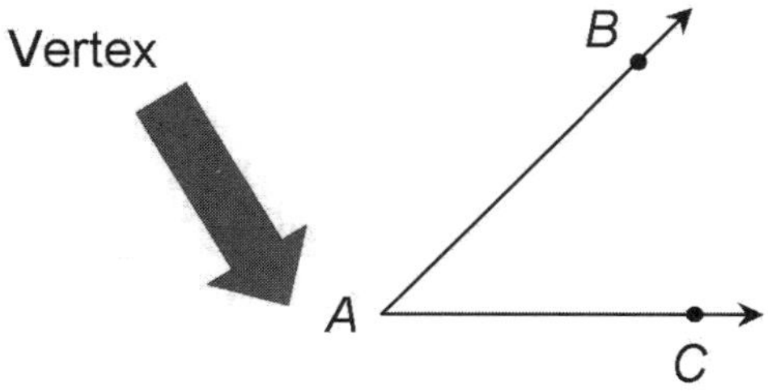

Vertex

This angle has vertex *D:*

Vertex

The plural of **vertex** is vertices.

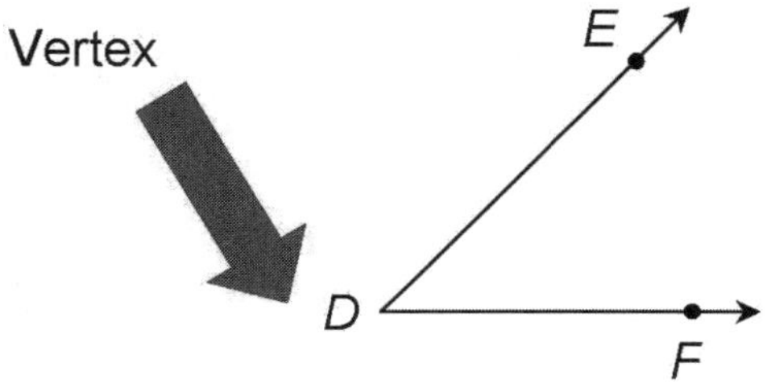

Naming angles

There is a special symbol for an angle in geometry.

The symbol is: ∠

Plane Geometry

There are three common ways to name angles.

1. Angles can be named by their lowercase letter inside the angle.

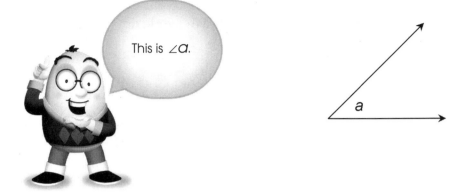

This is ∠*a*.

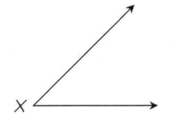

2. Angles can be named by their vertex. This is ∠X.

X

3. Angles can be named by their points. This is ∠*YXZ*.

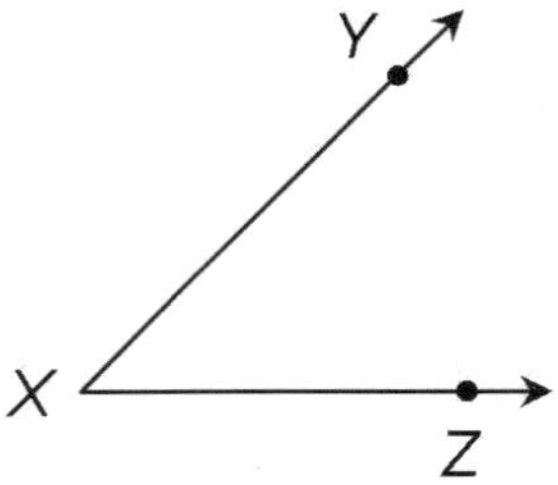

Y

X

Z

When we name an angle by three points, the vertex is always the middle letter.

This angle could be called ∠*BAC* or ∠*CAB*.

It would **not** be called ∠*ABC* or ∠*ACB*.

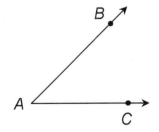

B

A

C

egghead's Guide to Geometry

Practice Questions

1. On the figure below, draw in four angles where lines *m* and *n* intersect. Label the angles *q, r, s,* and *t.*

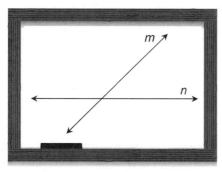

2. On the figure below, draw in four angles where lines *x* and *y* intersect. Label the angles *d, e, f,* and *g.*

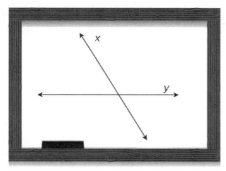

3. In the space below, draw an angle with ray *PQ*, ray *PR*, and vertex *P*.

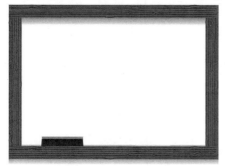

4. The following angle is angle *b*. Label the angle correctly.

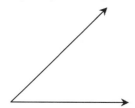

5. The following angle is ∠R. It also has ray *RT* and ray *RS*. Label the angle and the rays correctly.

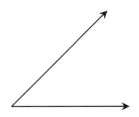

6. Write in the points to show that this is angle *LMN*.

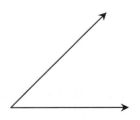

7. Write in three names for the angle shown. Use the ∠ symbol.

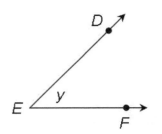

 **Solutions**

1.

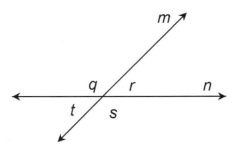

2.

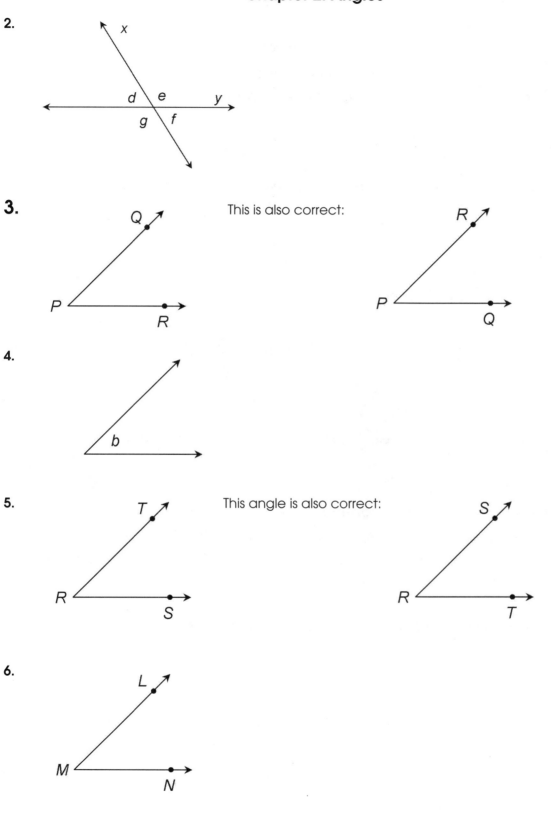

3.

This is also correct:

4.

5.

This angle is also correct:

6.

7.

∠y	Inside angle name
∠E	Vertex name
∠DEF or ∠FED	Three points name

Degrees

Now that we understand angle names and symbols, the next topic is angle size. Angles are measured in units called **degrees.**

Symbol

The symbol for degree in geometry is °.

We write 45 degrees as: 45°.

It's the same as the symbol for measuring temperature!

To write the measurement of an angle, we can use just the letter m for short. If we have ∠ABC that measures 45°, we would write m∠ABC = 45. It is not necessary to include the degree sign when using the short version.

Also, when writing the short version of an angle measurement, be sure to include the letter m. It is not correct to simply write ∠ABC = 45. The letter m must come before the angle sign.

Using a protractor

To measure angles in geometry, we use a tool called a **protractor**.

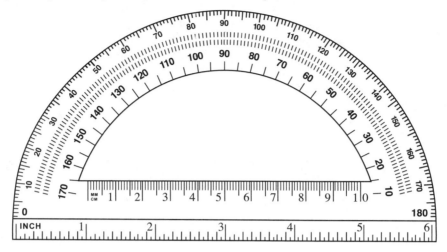

The protractor has a straight edge along the bottom. In the center of the straight edge, there is a dot or a small hole. This indicates the center point of the straight edge.

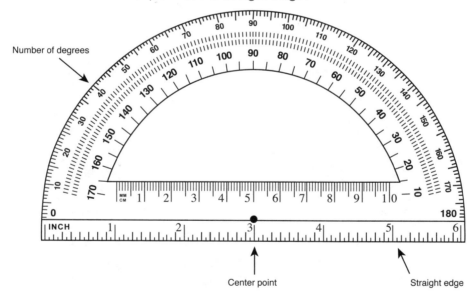

Plane Geometry

To use a protractor, place the center point on the vertex of the angle.

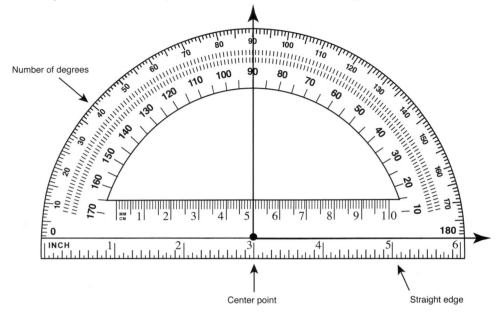

Align the bottom ray of the angle with the straight edge of the protractor. Note the number of degrees where the top ray crosses the protractor.

Examples

The degrees measure the size of the angle. The bigger the angle, the more degrees it has.

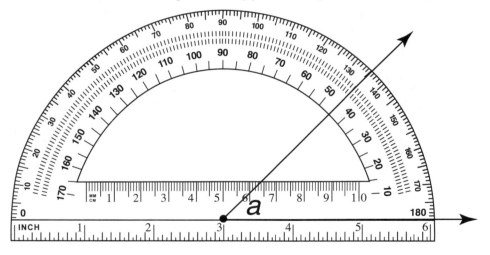

Angle *a* measures 45 degrees.

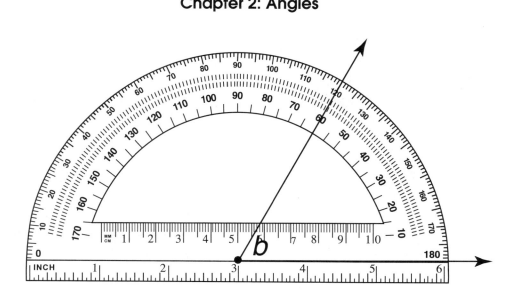

Angle *b* measures 60 degrees. It's a little bigger than ∠*a*.

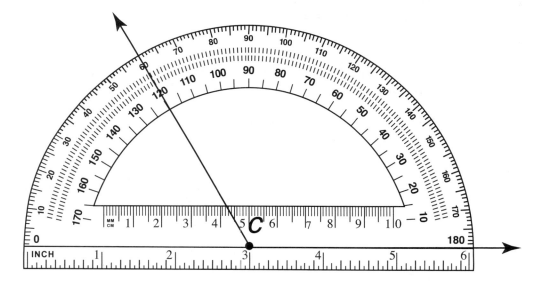

Angle *c* measures 120 degrees. The top ray slants backwards. If the angle opens to the right, use the degree measures on the lower part of the semi-circle, closest to the straight edge. If the angle opens to the left, use the degree measures at the top of the protractor, farthest from the straight edge.

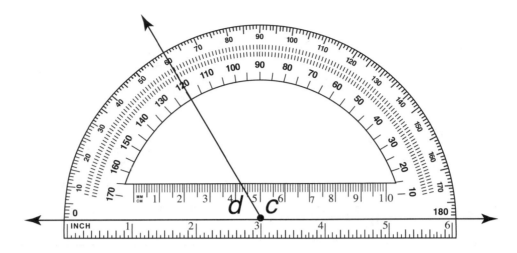

Angle *c* measures 120 degrees, while ∠*d* measures 60 degrees.

Practice Questions

Directions

Using a protractor, measure the following angles. Write the degree measurement inside each angle. Use the symbol for degrees.

If you don't have a protractor, estimate using the protractors shown.

1.

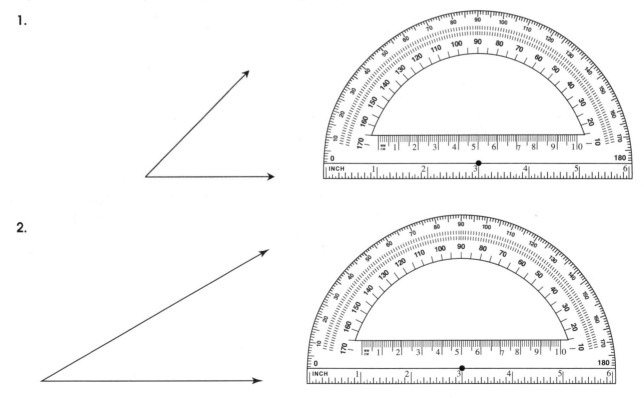

2.

3.

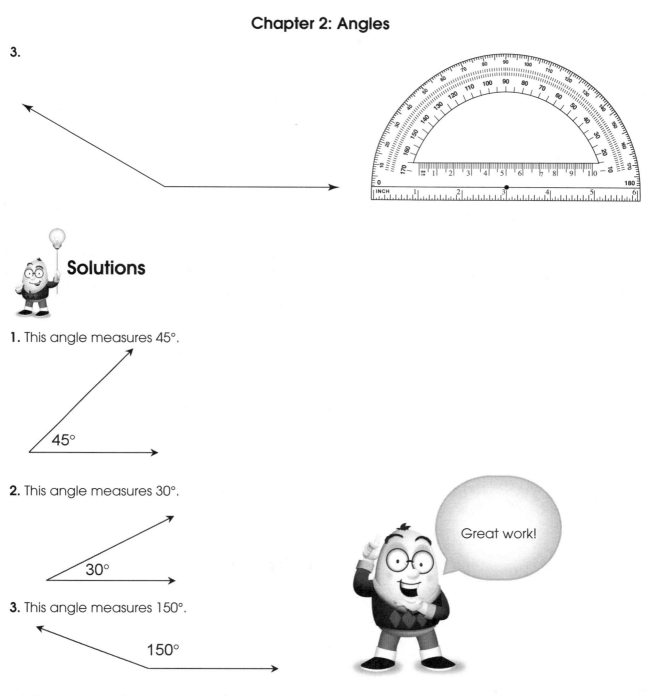

Solutions

1. This angle measures 45°.

45°

2. This angle measures 30°.

30°

3. This angle measures 150°.

150°

Great work!

Adding angle measures

You can add up the degree measures of angles just like you would normal numbers.

Examples

The measure of $\angle x$ is 60°. The measure of $\angle y$ is 12°. The measure of $\angle x$ plus the measure of $\angle y$ is 72°.

$$m\angle x + m\angle y = 72$$

In the figure shown, m∠R = 45 and m∠S = 60.

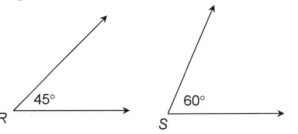

The measure of ∠R plus the measure of ∠S = 45 + 60, or 105.

Bisecting rays

In the last chapter, we learned about bisecting lines, or bisectors, which cross line segments at their midpoint. Angles can also have bisectors. A **bisecting ray** is a ray that divides one larger angle into two equal, smaller angles.

Examples

In the figure shown, ray *BJ* is a bisector. It divides ∠*ABC* into two equal angles.

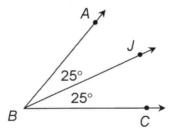

In this figure, ray *YO* bisects ∠*XYZ*. As the figure shows, m∠*XYO* = m∠*OYZ*.

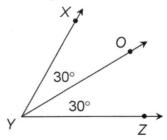

Practice Questions

1. What is m∠L + m∠K?

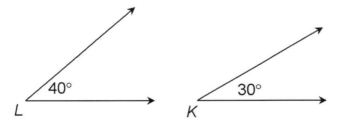

2. What is m∠ABC?

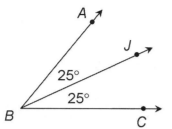

3. Ray *JA* bisects ∠HJK. What is the value of *x*?

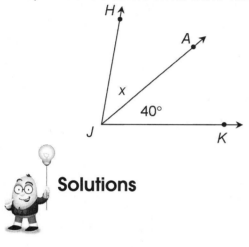

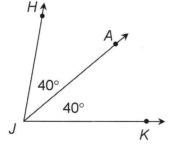

Solutions

1. The m∠L + m∠K = 40 + 30. The correct answer is 70.

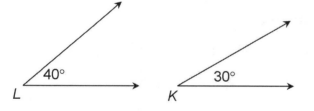

2. Angle *ABC* consists of two smaller angles, ∠ABJ and ∠JBC. To find m∠ABC, add m∠ABJ + m∠JBC:

$$m\angle ABJ + m\angle JBC = 25 + 25$$
$$= 50$$

3. Ray *JA* bisects ∠HJK. This means the two smaller angles are equal. Both angles measure 40°.

Degrees in a line

How do we know the number of degrees an angle has?

Let's start with the number of degrees in a line. A straight line measures 180°.

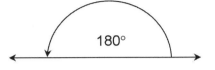

This is true for every straight line.

An angle that measures 180° is called a **straight angle.**

Supplementary angles

Say you take a straight line and divide it into two angles.

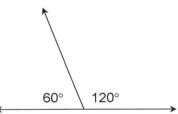

Now you have two angles that add up to 180°. There is a special name for these angles. They are called **supplementary angles.**

Supplementary angles are any two angles that add up to 180°.

Examples

These angles are supplementary angles:

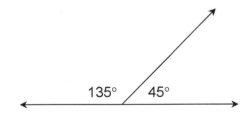

These angles are supplementary angles:

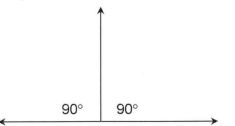

Chapter 2: Angles

Angles don't have to be on a line to be supplementary.

These angles are supplementary, too.

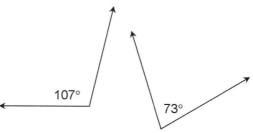

They just have to add up to 180°.

Practice Questions

1. In the figure below, ∠P and ∠R are supplementary. Write in the measure of ∠P.

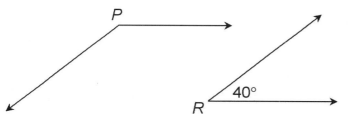

2. What is the measure of ∠XYW in the figure shown?

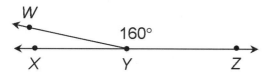

3. Which of the following angles are supplementary? Chose only two from the angles shown.

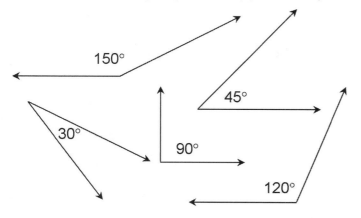

Solutions

1.

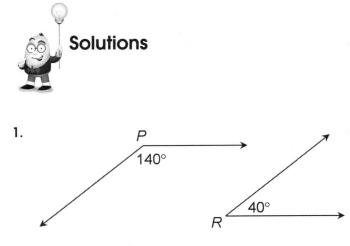

2. These two angles are supplementary. Their measures must add up to 180°. Subtract 160° from 180° to find the answer: 180° – 160° = 20°.

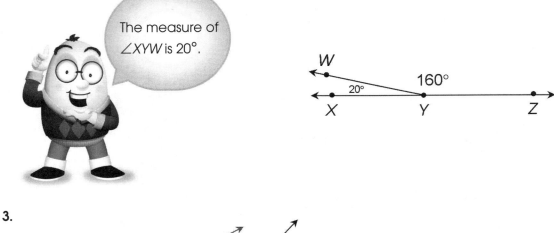

The measure of ∠XYW is 20°.

3.

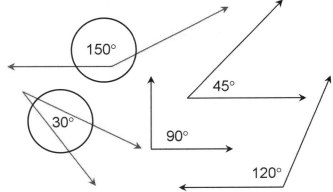

The two circled angles add up to 180°.

Right angles

If you take a straight line and divide it into two 90° angles, these angles are called **right angles.**
Right angles measure exactly 90°. As you can see, right angles are also supplementary.

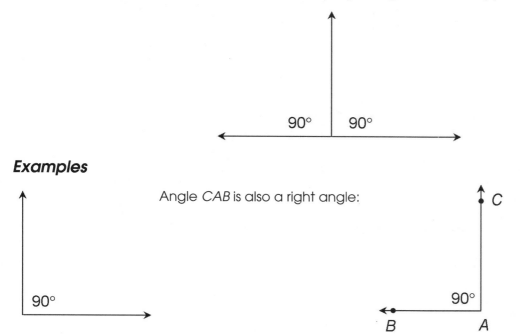

Examples

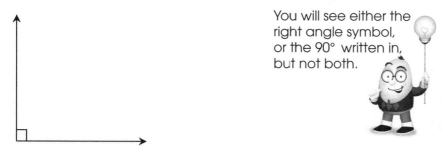

Angle *CAB* is also a right angle:

Symbol

Right angles have a symbol that is very commonly used. It might be the most frequently used
symbol in geometry.

The symbol is a small square on the inside of the angle. It looks like this:

You will see either the
right angle symbol,
or the 90° written in,
but not both.

When you see this symbol, you know you have a 90° angle, or right angle.

Complementary angles

If you take a right angle and divide it into two angles, you create another specific type of angle. These are called **complementary angles.**

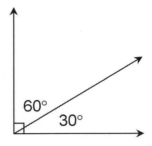

Complementary angles are two angles that add up to 90°.

Examples

Here are some examples of complementary angles:

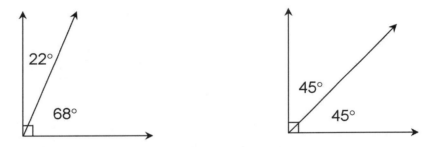

These are also complementary angles:

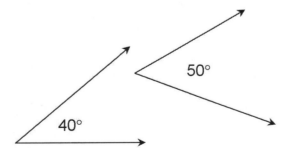

Two angles can be complementary even if they aren't within a right angle. As long as they add up to 90°, they are complementary.

Practice Questions

1. In the figure below, ∠K and ∠M are complementary. Write in the measure of ∠M.

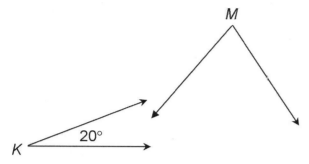

2. What is the measure of ∠CDE in the figure shown?

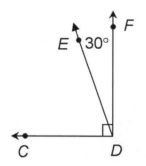

3. Which of the following angles are complementary? Circle the two correct angles.

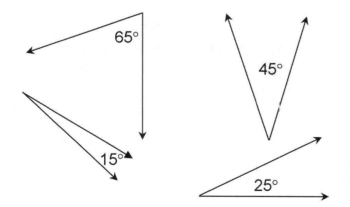

Solutions

1.

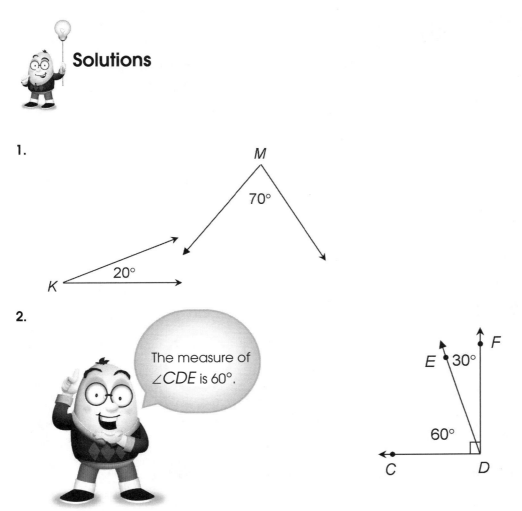

2.

The measure of ∠*CDE* is 60°.

This example clearly shows the complementary angles, which lie within the right angle. Again, the measure of these angles must add up to 90°. Subtract 30° from 90° to find the answer: 90° − 30° = 60°.

3.

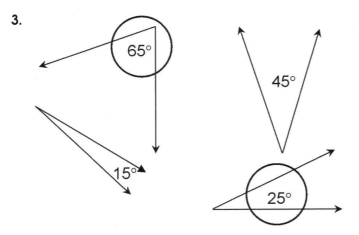

The two circled angles add up to 90°.

Perpendicular lines

While we're on the subject of right angles, here's one other important term. When two lines intersect at a right angle, we call the lines **perpendicular.**

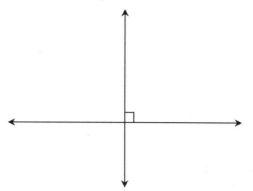

Perpendicular lines form four right angles:

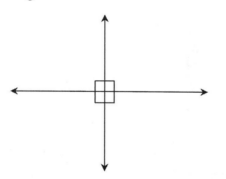

When you see perpendicular lines, you know all four angles are each 90°.

Symbol

In geometry, there are symbols for almost everything.

Perpendicular lines are no exception!

The symbol for perpendicular lines is: ⊥

Perpendicular lines x and y would be shown with the symbol like this: $x \perp y$.

This means that line *x* is perpendicular to line *y*.

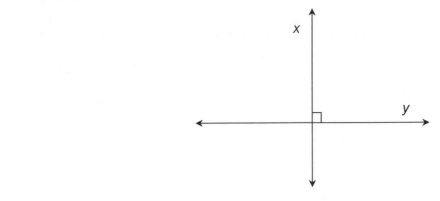

Practice Questions

1. Draw the symbol for the phrase line *t* is perpendicular to line *u*.

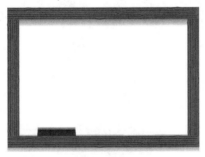

2. Draw two lines showing that line *t* is perpendicular to line *u*.

3. Draw two lines showing that line *h* ⊥ *j*.

4. Which of the following lines are perpendicular? Circle the correct answer found in the sets of lines shown.

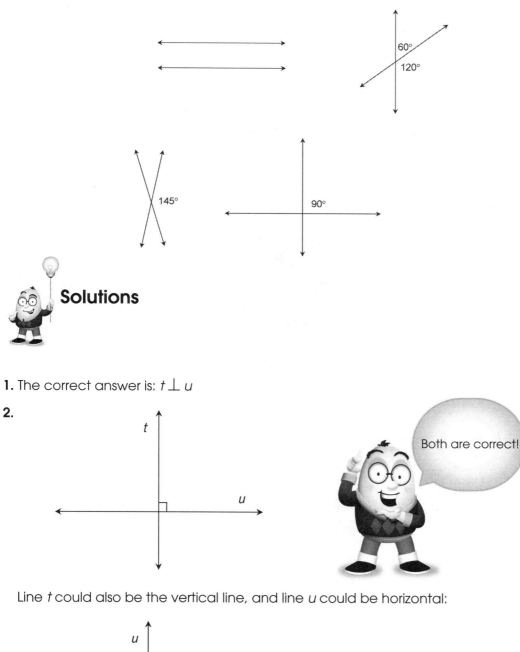

Solutions

1. The correct answer is: $t \perp u$

2.

Both are correct!

Line *t* could also be the vertical line, and line *u* could be horizontal:

3.

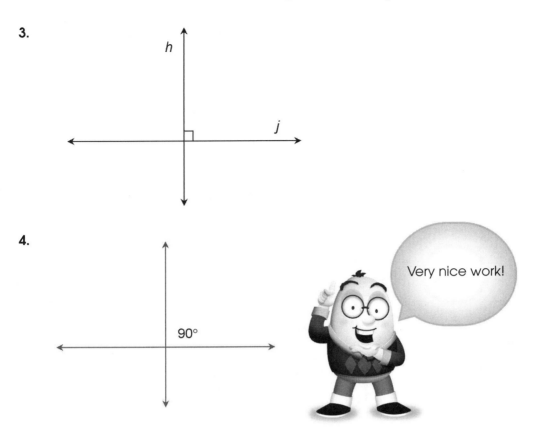

4.

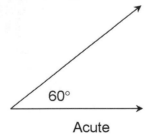

Very nice work!

The two lines intersect to form 90° angles.

Acute and obtuse angles

We are almost finished learning about angles. There are just a few more to know.

Angles that are less than 90° are called **acute angles.**

60°

Acute

Angles that are less than 180°, but greater than 90°, are called **obtuse angles.**

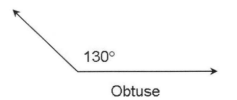

130°

Obtuse

Examples

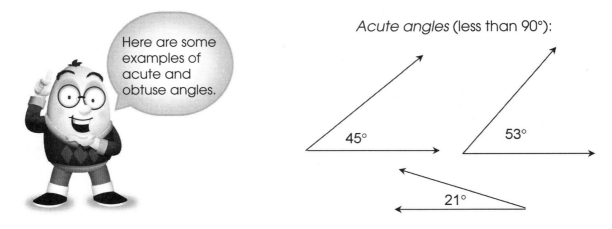

Acute angles (less than 90°):

Obtuse angles (greater than 90°, but less than 180°):

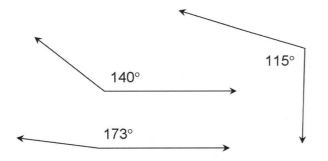

Practice Questions

1. Which of the angles shown below are acute angles? Circle all that apply.

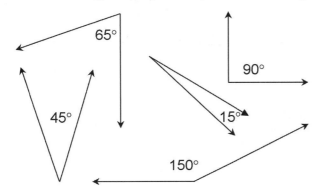

2. Which of the angles shown below are obtuse angles? Circle all that apply.

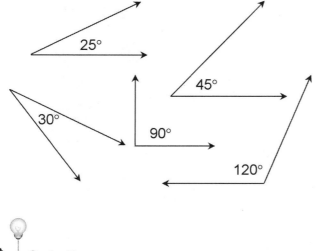

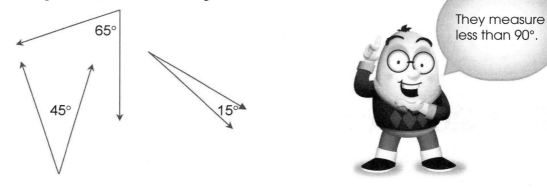

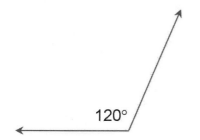

💡 **Solutions**

1. The angles shown are acute angles.

They measure less than 90°.

2. There is only one obtuse angle in this group. It is the angle shown. It measures more than 90°, but less than 180°.

120°

Adjacent angles

Angles that lie next to each other are known as **adjacent angles.** To be adjacent, angles must share one side and have the same vertex.

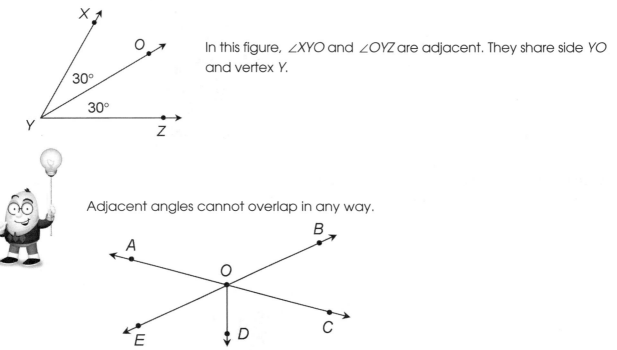

In this figure, ∠XYO and ∠OYZ are adjacent. They share side YO and vertex Y.

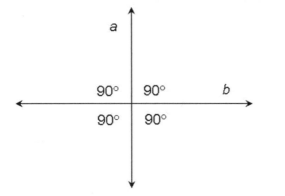

Adjacent angles cannot overlap in any way.

In the figure above, ∠BOC and ∠COD are adjacent. They share side OC and vertex O. However, ∠BOC and ∠BOD are not adjacent, because ∠BOC is part of ∠BOD. Their interior spaces overlap.

Adjacent angles can be either acute or obtuse. It's also possible to have adjacent right angles. The important thing is to know that **adjacent** means next to each other.

Vertical angles

Where perpendicular lines cross, four equal angles are formed.

All of the angles measure 90°.

Plane Geometry

But what about lines that aren't perpendicular? When they intersect, do equal angles form?

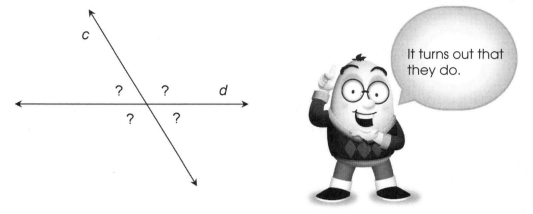

It turns out that they do.

Each pair of intersecting lines forms some equal angles. These are called **vertical angles.** Vertical angles are the angles that fall opposite each other when the lines cross.

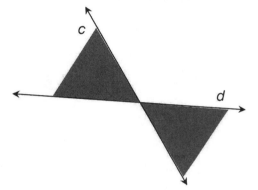

The shaded areas show one pair of vertical angles. The shaded angles are equal in size. The unshaded angles are also equal. They are the second vertical pair.

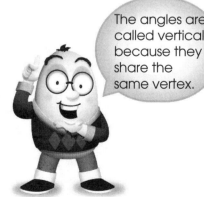

The angles are called vertical because they share the same vertex.

Examples

In the figure shown, *a* and *a* are vertical angles. They are opposite each other when the two lines cross. They measure the same number of degrees.

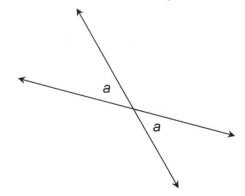

In the figure shown, *b* and *b* are vertical angles. They are equal, opposite angles.

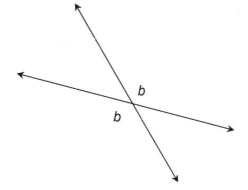

This figure has two pairs of vertical angles. The angles marked *a* are equal. The *b* angles are equal too.

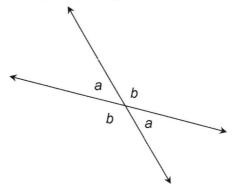

Congruence

In geometry, when two angles are equal, we say they are **congruent.**

Congruence is **VERY** important in geometry. Many questions ask about congruent lines, angles, and shapes. There are even some special symbols to show congruence.

When we want to write that two angles are congruent, we use this symbol:

$$\cong$$

Congruent angles *S* and *T* would be shown with the symbol like this:

$$\angle S \cong \angle T$$

This means that angle *S* is congruent to angle *T*. The two angles are equal.

If we want to **show** that two angles are equal, for instance in a figure, we would use some special marks:

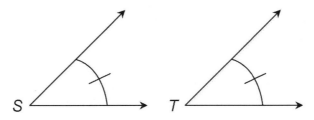

The lines in the figure show that angle *S* is equal to angle *T*. The half-circle is called an **arc.** The small lines on the arcs are called **hash marks.**

If we have more than one pair of equal angles, we use single hash marks for one pair and double hash marks for the other:

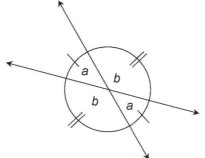

We can even use triple hash marks!

It doesn't matter how many hash marks you use to show congruence (within reason, of course). You just have to use the same number for each item in the pair.

Those angles are definitely congruent!

Practice Questions

1. In the figure below, show that the vertical angles are congruent. Use single hash marks for one pair, and double hash marks for the other.

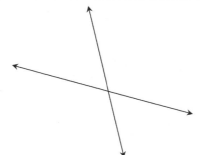

2. In the figure below, which angles are vertical angles? Identify two pairs.

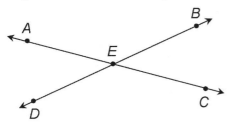

3. In the figure below, what is the measure of ∠*r*?

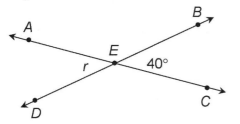

4. This question is a little bit tougher. In the figure below, what is the measure of ∠*s*?

Solutions

1.

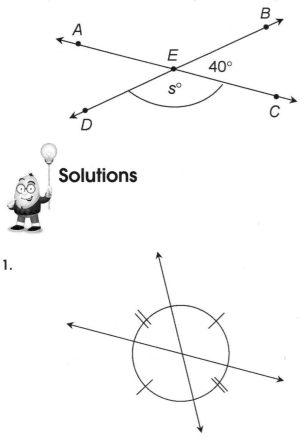

2. There are two sets of vertical angles here. You could have selected ∠AED and ∠BEC. The other vertical angles are ∠AEB and ∠DEC.

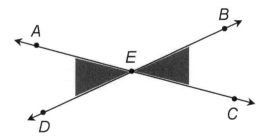

3. Angle *r* lies opposite an angle that measures 40°. The angles are vertical angles, so they are equal. Angle *r* also measures 40°.

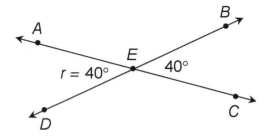

4. Angle *s* lies on a straight line with another angle that measures 40°. The angles are supplementary angles, so they add up to 180°. Subtract 40° from 180° to find the measure of ∠*s*:

$$180° - 40° = 140°$$

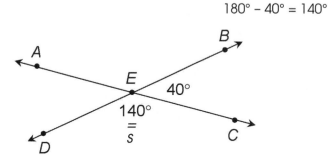

Corresponding angles

There is one final important angle type to know about. These are **corresponding angles.**

Remember in the chapter about lines, we talked about transversals?

Chapter 2: Angles

When a transversal crosses two parallel lines, the lines look like this:

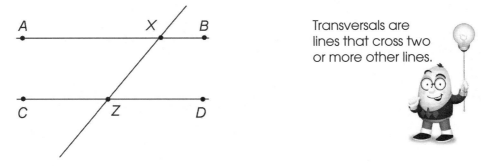

Transversals are lines that cross two or more other lines.

Eight different angles are created. Four are at the top, and four are at the bottom. Some of the top angles match the bottom angles. These are called **corresponding angles.**

Examples

For instance, these two angles are congruent. They are corresponding angles:

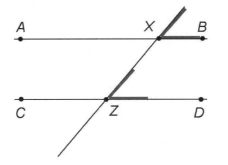

These two angles are also congruent. They are corresponding angles:

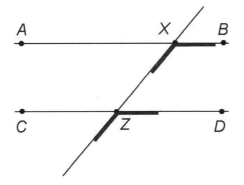

Plane Geometry

There are four pairs of corresponding angles total:

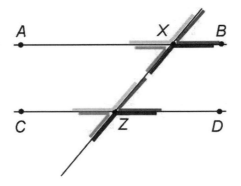

The angles on the top line match the angles in the same location on the bottom line.

Practice Questions

1. In the figure below, ∠r and ∠s are corresponding angles. $\overline{AB}$ is parallel to $\overline{CD}$. Draw ∠s in its correct location.

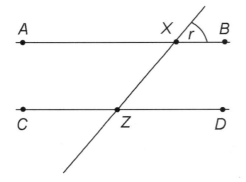

2. In the figure shown, $\overline{WX}$ is parallel to $\overline{YZ}$, and ∠a measures 60 degrees. What is the measure of ∠b?

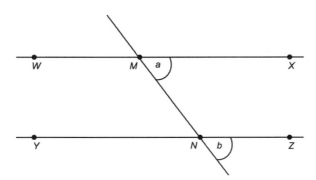

Chapter 2: Angles

3. In the figure below, which angles are corresponding angles? Which angles are vertical angles?

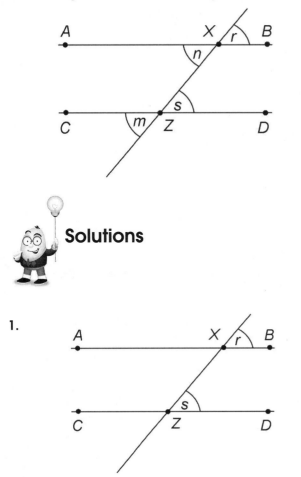

![lightbulb] **Solutions**

1.

2. The measure of ∠b is 60°. The figure contains two parallel lines crossed by a transversal. Angles *a* and *b* are corresponding angles, which means they're congruent. The measurement of ∠b is equal to the measurement of ∠a, which is 60°.

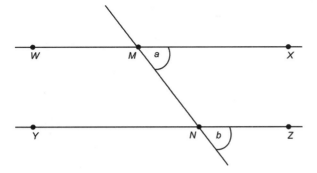

3. In the figure, ∠*r* and ∠*s* are corresponding angles. Angles *n* and *m* are corresponding angles. Angles *n* and *r* are vertical angles. Angles *m* and *s* are vertical angles, too.

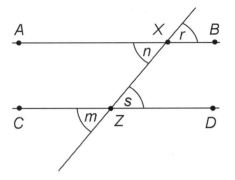

Chapter Review

1. On the figure below, draw in four angles where lines *r* and *s* intersect. Label the angles *t, u, v,* and *w*.

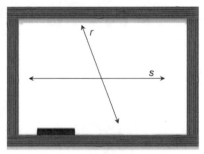

2. In the space below, draw an angle with ray *YX*, ray *YZ*, and vertex *Y*.

3. What is the measure of ∠*c* in the figure shown?

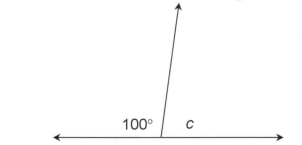

100°   *c*

4. In the figure below, ∠F and ∠G are complementary angles. Write in the measure of ∠G.

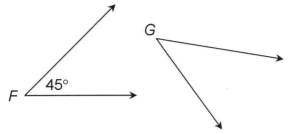

45°

5. In the figure below, which angles are corresponding angles? Which angles are vertical angles?

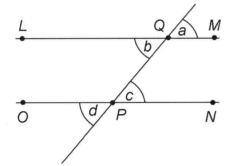

6. On the map below, West Avenue crosses Main Street as shown. Which of the buildings lies on a corner that forms a vertical angle with the corner for building B?

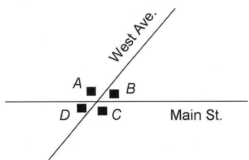

7. The figure below shows the slices of pie that each person in the Smith family ate for dessert.

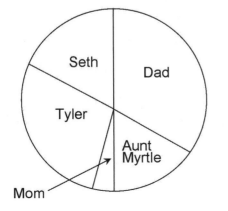

Which two people ate pie slices that formed adjacent angles?

A) Seth and Aunt Myrtle

B) Mom and Dad

C) Tyler and Dad

D) Mom and Tyler

8. Which of the following angles are supplementary? Circle two of the angles shown.

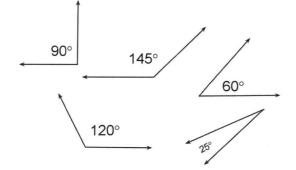

9. In the figure below, $\angle x$ and $\angle y$ are corresponding angles. Line segment *LM* is parallel to line segment *NO*. Draw in $\angle y$ in its correct location.

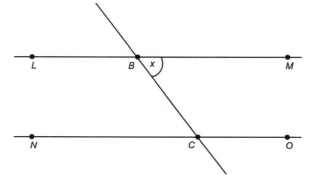

10. On the map shown, Pine Street is parallel to Spruce Street. Park Avenue intersects both Pine and Spruce.

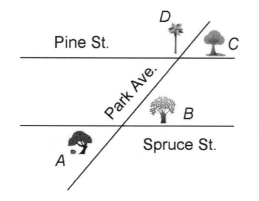

Which tree is located on a corner that forms a corresponding angle with the corner of oak tree C?

A) Apple tree A

B) Locust tree B

C) Oak tree C

D) Palm tree D

Chapter 2: Angles

11. Which of the following angles are complementary? Circle two of the angles shown.

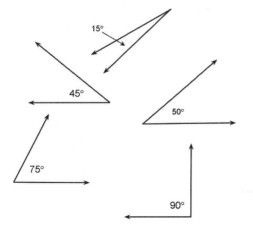

12. A plank of wood is placed over a fulcrum to form a seesaw, as shown in the figure. The bottom of the plank forms a straight angle. What is the value of *x*?

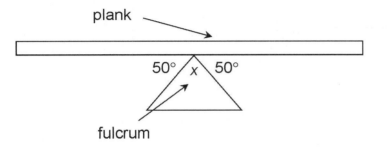

13. In the figure below, which angles are corresponding angles? Which angles are vertical angles?

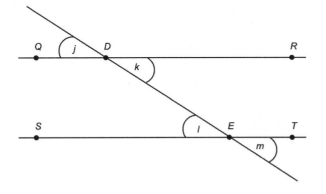

14. In the figure shown, the back leg of a chair is parallel to the front leg. If m∠s = 90, what is the m∠f?

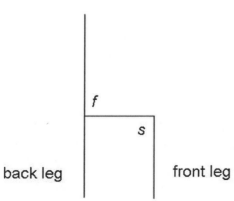

15. In the figure shown, line segment *AB* is parallel to line segment *CD* and ∠u measures 45 degrees. What is the measure of ∠v?

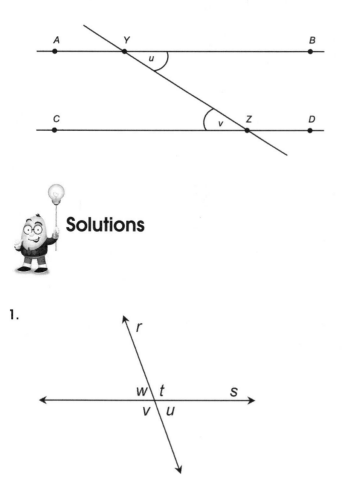

Solutions

1.

2. One correct answer is shown below.

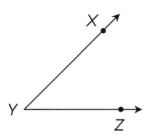

3. The two angles are supplementary. They lie on a straight line, so their measures add up to 180°. Subtract to find the measure of *c*: 180° − 100° = 80°.

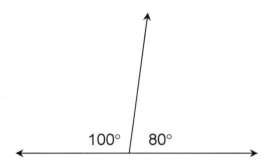

4. The correct answer is shown below. Since ∠F and ∠G are complementary angles, their measures add up to 90°.

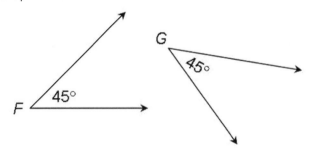

5. In the figure, ∠*a* and ∠*c* are corresponding angles. Angles *b* and *d* are corresponding angles. Angles *a* and *b* are vertical angles. Angles *c* and *d* are vertical angles, too.

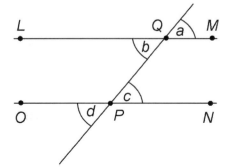

Plane Geometry

6. The correct answer is building D. The corner for building D lies directly across from the corner on which building B is located. Buildings A and C are also located on corners that are vertical angles.

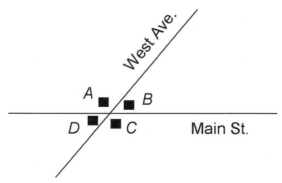

7. The correct answer is D, Mom and Tyler. The dotted line in the figure shows the side shared by the two angles. The angles also share a vertex, which is the point where the lines cross in the center of the pie.

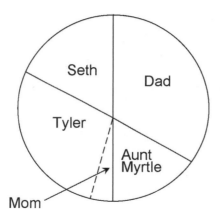

8. The correct answer is shown below.

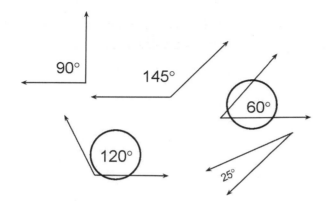

The two circled angles add up to 180°.

9. The correct answer is shown below.

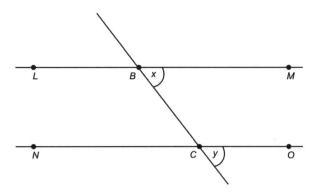

10. The correct answer is B, locust tree. Oak tree C is located on a corner that forms a corresponding angle with the corner of locust tree B. Locust tree B and apple tree A are located on corners that form vertical angles.

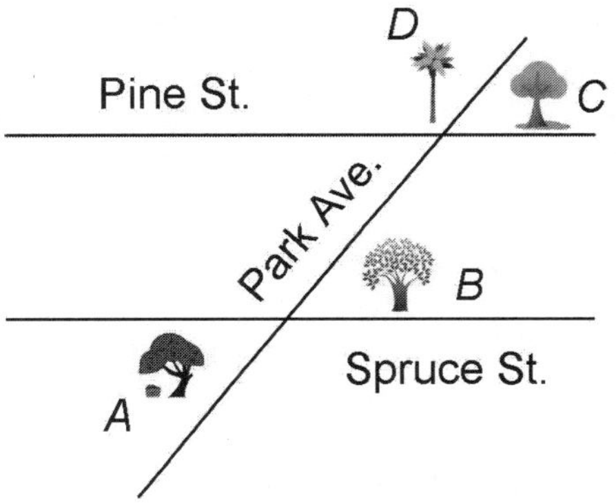

11. The correct answer is shown below.

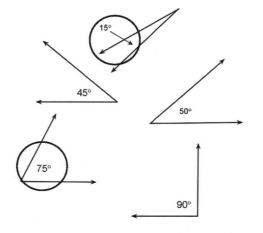

The two circled angles add up to 90°.

12. The value of *x* is 80°.

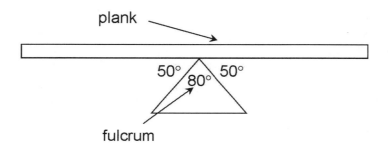

The bottom of the plank forms a straight angle. A straight angle measures 180°, so the three angles shown must add up to 180°. To find the value of *x*, subtract the two 50° angles from 180°:

$$180° - 50° - 50° = 80°$$

13. In the figure, ∠*j* and ∠*l* are corresponding angles. Angles *k* and *m* are corresponding angles. Angles *j* and *k* are vertical angles. Angles *l* and *m* are also vertical angles.

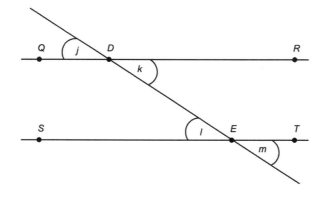

14. The m∠f = 90°.

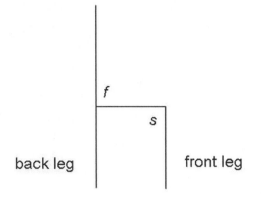

The front and back legs of the chair are parallel. The seat of the chair forms a transversal that crosses the front and back legs. The seat of the chair forms a 90° angle, ∠s, with the front leg.

If the seat of the chair and the front leg form a 90° angle, that means they are perpendicular. The front and back legs are parallel, so the seat is perpendicular to the back leg also. The measure of ∠f is therefore 90°.

15. The measure of ∠v is 45°. The figure contains two parallel lines crossed by a transversal. Angle u has a vertical angle, circled, which is equal in measure to ∠u. The circled angle therefore measures 45°.

The circled angle and ∠v are corresponding angles, which means they're congruent. The measurement of ∠v is therefore also 45°.

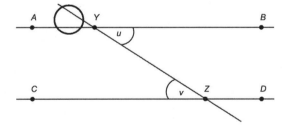

Chapter 3

Quadrilaterals

Hi! I'm egghead. I will teach the following concepts in this chapter:

What is a quadrilateral?
Naming quadrilaterals
Main types of quadrilaterals
Rectangles
Parallel and equal sides
Finding perimeter
Finding area
Squares
Finding perimeter
Finding area
Parallelograms

Finding angles and length
Finding area
Base and height
Trapezoids
Finding perimeter
Finding area
Base and height
Rhombuses
Finding perimeter
Finding area
Base and height

What is a quadrilateral?

A **quadrilateral** is any closed, two-dimensional shape with four sides that are straight lines.

In Latin, the root
quad means four.

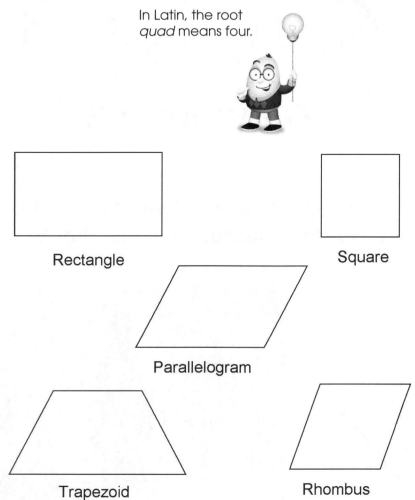

Rectangle

Square

Parallelogram

Trapezoid

Rhombus

There are also four-sided figures with random shapes:

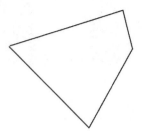

For the most part, geometry test questions ask about quadrilaterals with recognizable shapes.

We'll focus on these shapes in this chapter.

Naming quadrilaterals

To name a quadrilateral, use its points. The points are also called **vertices.**

Every quadrilateral has four vertices. So, it will have four letters in its name.

Examples

This is quadrilateral *ABCD*:

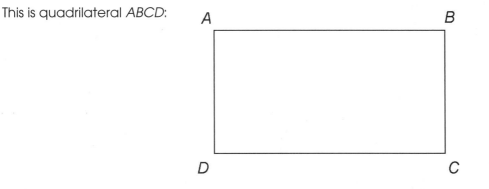

The points can be listed in clockwise or counterclockwise order.

This is quadrilateral *WXYZ*:

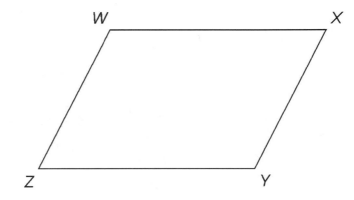

It could also be called quadrilateral *WZYX*.

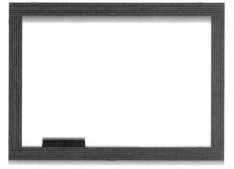

Practice Questions

1. In the space below, draw quadrilateral *MNOP*.

2. What is the name of this quadrilateral?

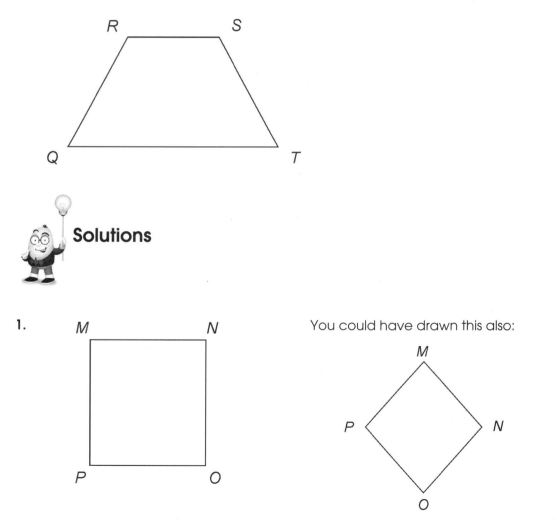

Solutions

1.

M   N

P   O

You could have drawn this also:

M

P   N

O

A four-sided figure can have many different shapes!

egghead's Guide to Geometry

2.

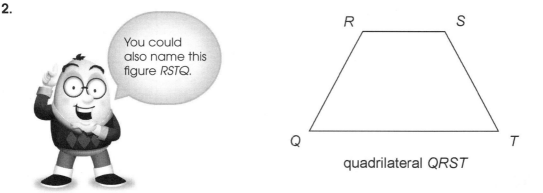

quadrilateral *QRST*

Main types of quadrilaterials

There are several types of four-sided figures, as we saw with the examples above.

These are:

1. Rectangles

2. Squares

3. Parallelograms

4. Trapezoids

5. Rhombuses (or rhombi)

Rectangles

Rectangles are four-sided figures with four right angles.

Examples

This is rectangle *ABCD*:

This is rectangle *HIJK*:

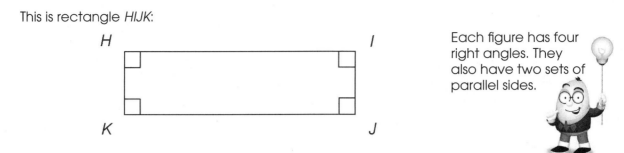

Each figure has four right angles. They also have two sets of parallel sides.

Parallel and equal sides

Every rectangle has opposite sides that are parallel.

In the figure below, side *HI* is parallel to side *KJ*.

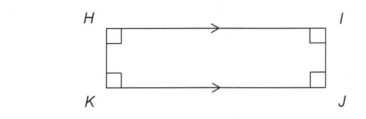

The arrows show the lines are parallel.

Side *KH* is parallel to side *JI*.

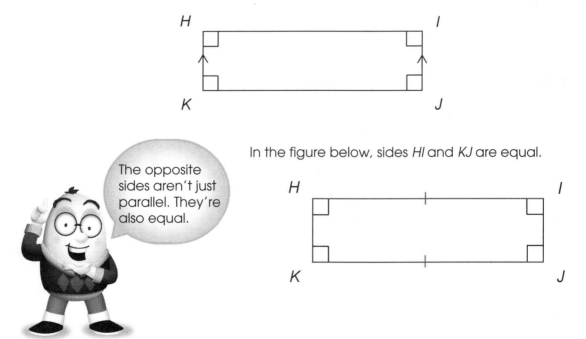

The opposite sides aren't just parallel. They're also equal.

In the figure below, sides *HI* and *KJ* are equal.

Sides *KH* and *JI* are equal, too.

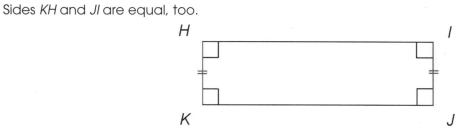

Finding perimeter

Perimeter is the distance around the outside of a shape. The dotted arrows show the perimeter of the rectangle.

The perimeter of a rectangle is the sum of the lengths of its four sides.

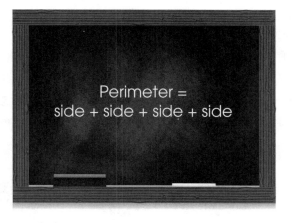

Perimeter =
side + side + side + side

Examples

The perimeter of rectangle *CDFE* is 5 + 3 + 5 + 3.

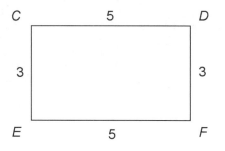

The perimeter is 16.

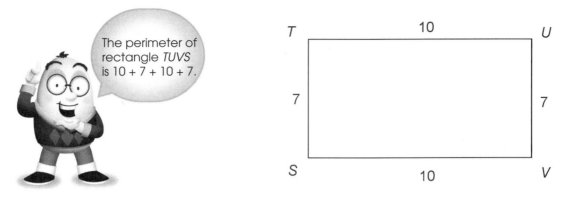

The perimeter of rectangle *TUVS* is 10 + 7 + 10 + 7.

The perimeter is 34.

The perimeter of rectangle *UVWX* is 15 + 20 + 15 + 20.

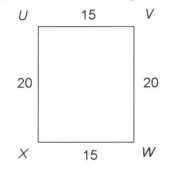

The perimeter is 70.

Formula

To find measurements like perimeter, it's helpful to know the formula. Since the perimeter is the sum of the lengths of the sides of a shape, to find the perimeter of a rectangle, we use the formula Perimeter = side + side + side + side. We can abbreviate perimeter as *P*, and side as *s*. The formula looks like this:

$$P = s + s + s + s$$

Chapter 3: Quadrilaterals

The sides of a rectangle have special names.

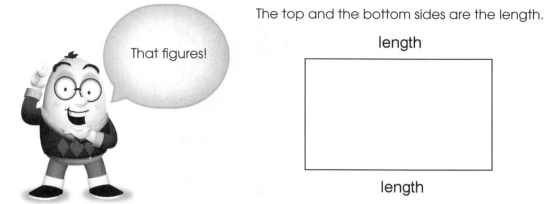

That figures!

The top and the bottom sides are the length.

length

length

The right and the left sides are the width.

width

width

Another way to write the perimeter formula is like this:

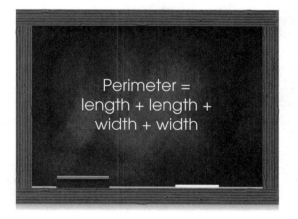

Perimeter =
length + length +
width + width

Plane Geometry

Since there are two lengths and two widths, the formula can also be written this way:

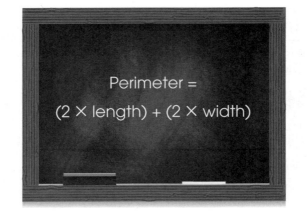

Perimeter =
(2 × length) + (2 × width)

If we abbreviate, this becomes:

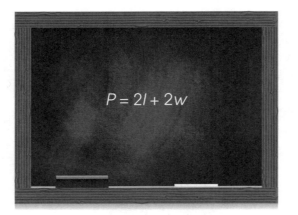

$P = 2l + 2w$

There are many ways to write the perimeter formula. Use whichever way works best for you.

Practice Questions

1. What is the perimeter of rectangle *FGHJ*?

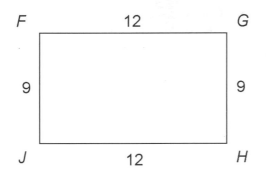

2. What is the perimeter of rectangle *QRST*?

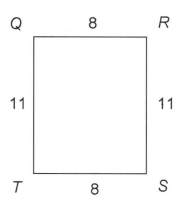

3. What is the perimeter of rectangle *LMNO*?

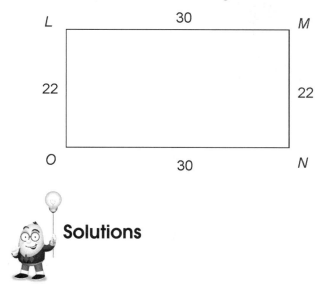

Solutions

1. To find the perimeter, add up the lengths of the four sides. The perimeter is 12 + 9 + 12 + 9 = 42.

What other way can you find the perimeter of this rectangle?

2. The perimeter is 8 + 11 + 8 + 11 = 38.

3. The perimeter is 30 + 22 + 30 + 22 = 104.

Finding area

The **area** of a shape is the total amount of space a shape covers. The shading in the figure shows the area of the rectangle.

To find the area of a rectangle, multiply its length times its width. As we saw in Chapter 1, the area of any figure is always given in square units. If the measurements of the figure are in inches, then the area will be in square inches. If the measurements are in centimeters, the area is in square centimeters.

Examples

To find the area of rectangle *CDFE*, multiply the length (5) times the width (3).

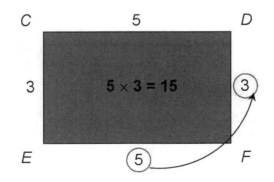

The area is 15 square units.

To find the area of rectangle *TUVS*, multiply the length (10) times the width (7).

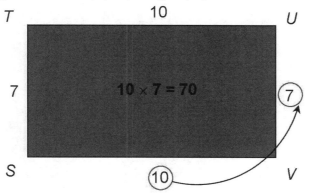

The area is 70 square units.

To find the area of rectangle *UVWX*, multiply the length (15) times the width (20).

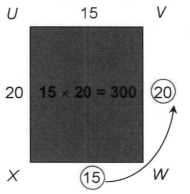

The area is 300 square units.

Formula

The formula for the area of a rectangle is:

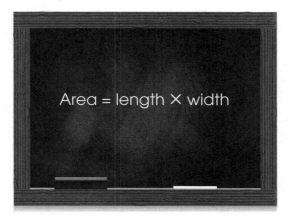

Area = length × width

Plane Geometry

If you want to abbreviate, you can write the formula like this:

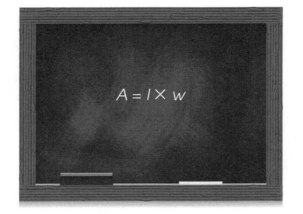

$A = l \times w$

Practice Questions

1. What is the area of rectangle *FGHJ*?

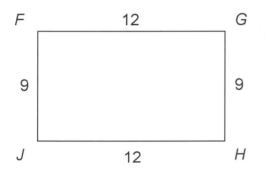

2. What is the area of rectangle *QRST*?

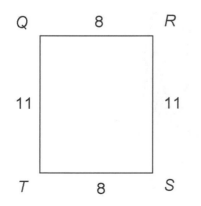

Solutions

1. The area is 108 square units.

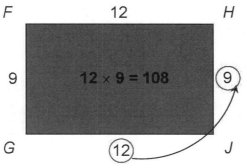

2. The area is 88 square units.

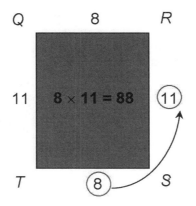

Squares

Squares are a lot like rectangles. They are both four-sided figures with four right angles.

In fact, squares *ARE* rectangles! A square is just a rectangle with four equal sides.

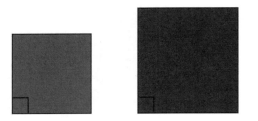

A square has four right angles and four equal sides. If a square is a rectangle, then its opposite sides must be what? That's right—parallel.

Examples

This square is labeled *DEFG*.

This square has hash marks showing four equal sides:

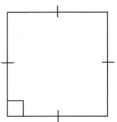

This square shows the length of each side. Side *JK* is parallel to side *ML*, and side *MJ* is parallel to side *LK*.

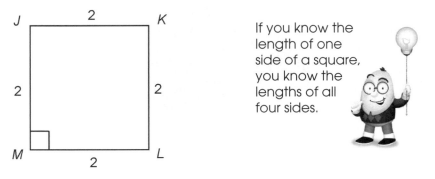

If you know the length of one side of a square, you know the lengths of all four sides.

Finding perimeter

The perimeter of a square is the sum of the lengths of its sides:

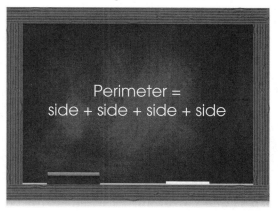

Since all four sides are equal, we could also multiply the length of one side by 4 to get the perimeter of the square:

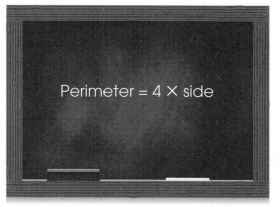

Formula

To shorten the formula, we write it like this:

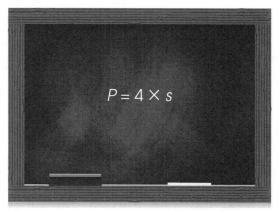

Practice Questions

1. Find the perimeter of square *JKLM*.

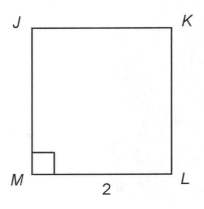

2. Find the perimeter of square *BCDE*.

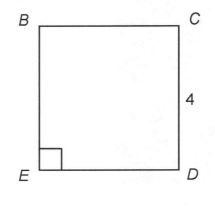

 **Solutions**

1. The perimeter is 2 + 2 + 2 + 2, which equals 8.

2. To find the perimeter, add together the lengths of all four sides. The perimeter is 4 + 4 + 4 + 4, which equals 16.

Finding area

To find the area of a square, we multiply its length times its width.

The same as for a rectangle!

However, since the length and width of a square are equal, we multiply one side by another:

Area = side × side

Formula

To shorten the area formula, we write:

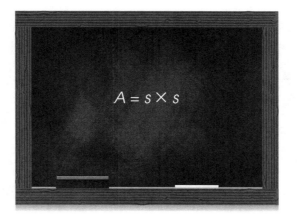

$$A = s \times s$$

Practice Questions

1. What is the area of square *JKLM*?

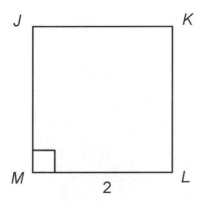

2. What is the area of square *BCDE*?

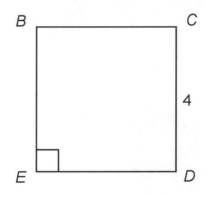

3. What is the area of square *RSTU*?

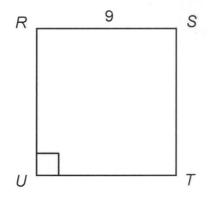

 **Solutions**

1. To find the area, multiply two sides.

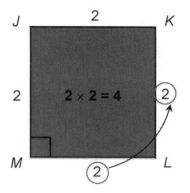

The area of square *JKLM* is 4 square units.

2. To find the area, multiply two sides.

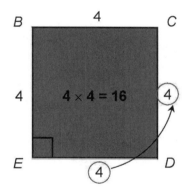

The area of square *BCDE* is 16 square units.

3. To find the area, multiply two sides.

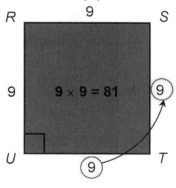

The area of square *RSTU* is 81 square units.

Good work!

Parallelograms

Another common quadrilateral in geometry is the parallelogram. **Parallelograms** are four-sided shapes with opposite parallel sides.

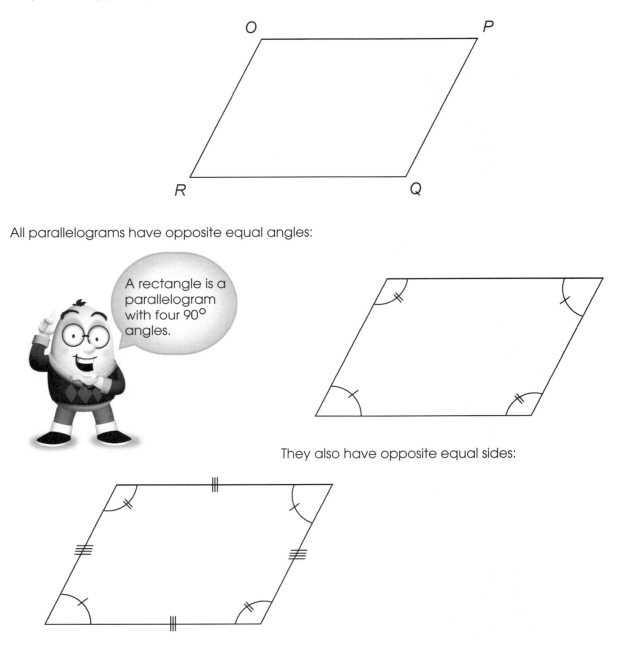

All parallelograms have opposite equal angles:

A rectangle is a parallelogram with four 90° angles.

They also have opposite equal sides:

Let's look at some examples.

Examples

This is parallelogram *WXYZ*. Angle *Z* measures 60°.

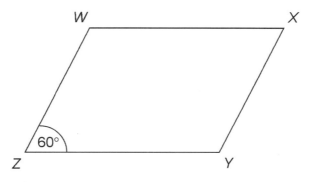

This is parallelogram *EFGH*. Side *EF* has length 14.

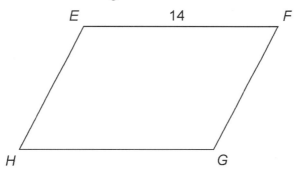

This is parallelogram *OPQR*. Sides *OP* and *RQ* have length 5. Sides *RO* and *QP* have length 4.

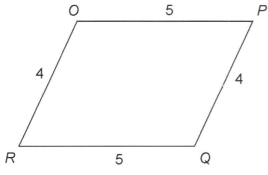

Opposite sides are parallel and equal.

Finding angles and length

Using what you know about parallelograms, answer the questions below.

Practice Questions

1. What is the measure of ∠X in the parallelogram shown?

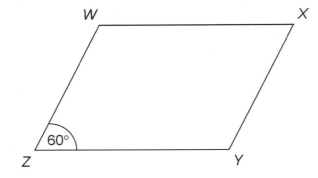

2. What is the length of side *HG* in the parallelogram shown?

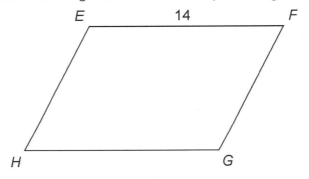

3. Quadrilateral *QRST* is a parallelogram. What are the lengths of sides $\overline{ST}$ and $\overline{TQ}$?

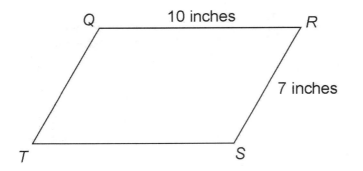

4. Parallelogram *UVST* is shown in the figure. What is the value of *x*?

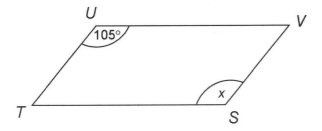

egghead's Guide to Geometry

Solutions

1. Parallelograms have equal opposite angles. This means ∠Z is congruent to ∠X.

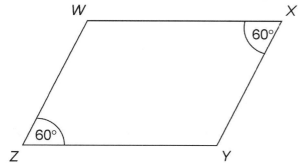

2. Parallelograms have equal opposite sides. This means $\overline{EF}$ is congruent to $\overline{HG}$.

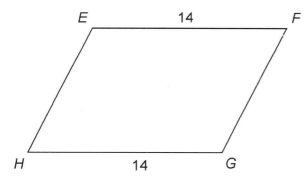

3. The length of side $\overline{ST}$ is 10 inches, and the length of side $\overline{TQ}$ is 7 inches.

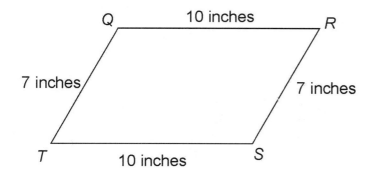

Since QRST is a parallelogram, we know that $\overline{QR}$ and $\overline{ST}$ are congruent. So, the measure of $\overline{ST}$ is 10 inches. Sides $\overline{RS}$ and $\overline{TQ}$ are also congruent, so the measure of $\overline{TQ}$ must be 7 inches.

4. The value of *x* is 105°.

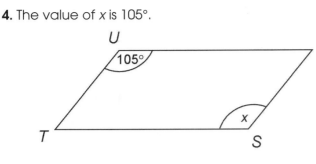

In parallelogram *UVST*, ∠*U* and ∠*S* are congruent. Therefore, *x* also equals 105°.

Perimeter questions

Parallelogram questions often ask about perimeter. You can find the perimeter of a parallelogram by adding up the lengths of the sides, just like you would do with a rectangle or square.

Practice Questions

1. Find the perimeter of parallelogram *WXYZ*.

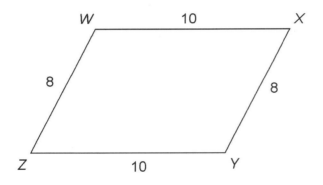

2. Find the perimeter of parallelogram *OPQR*.

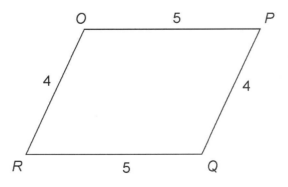

3. This one is a little more difficult. What is the perimeter of parallelogram *EFGH*?

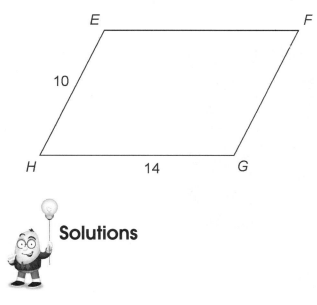

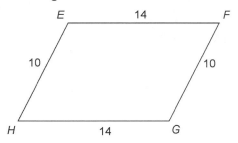

Solutions

1. The perimeter is equal to 10 + 8 + 10 + 8. The correct answer is 36.

2. The perimeter is equal to 5 + 4 + 5 + 4. The correct answer is 18.

3. We know that side *HE* has length 10. Side *HE* is equal to side *GF*. So, side *GF* measures 10 also. The remaining side measures 14.

Now add the lengths of the sides to find the perimeter:

$$14 + 10 + 14 + 10 = 48$$

Finding area

Finding the area of a parallelogram is similar to finding the area of a rectangle. Instead of multiplying length × width, we use base × height.

Formula

The formula for the area of a parallelogram can be written this way:

Area = base × height

It can also be written in shorthand:

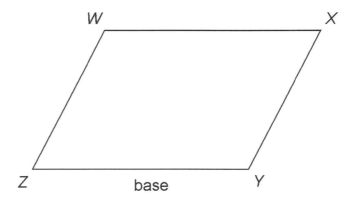

$A = b \times h$

That's geometry code for area!

Base and height

The **base** of a parallelogram is the side on the bottom.

The height of the parallelogram is a little harder to find.

It is NOT the length of the slanted side.

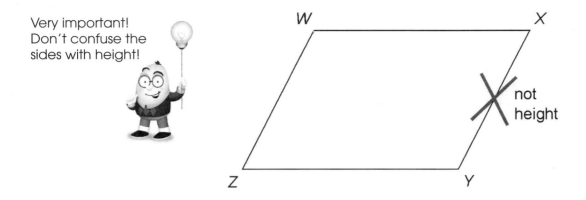

Very important! Don't confuse the sides with height!

The **height** of a rectangle is the length from the top side to the bottom side. The dotted line shows the height of parallelogram *WXYZ*:

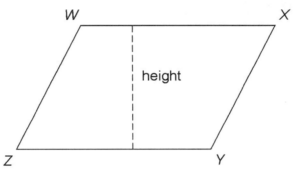

Sides *XY* and *WZ* are a little longer than the dotted height line. The sides are slanted, so the distance between their endpoints is longer.

The height of the parallelogram might also extend straight down from point *X*. The dotted line on the right shows a second height line:

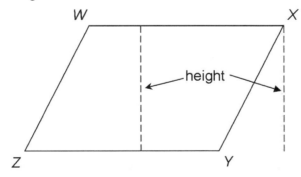

The dotted lines are equal. To show the height correctly, draw the height line straight down from the top side of the parallelogram to the bottom side. It cannot be slanted.

Plane Geometry

Some figures show an extra line extending out from the bottom side. The solid line extends out from point Y:

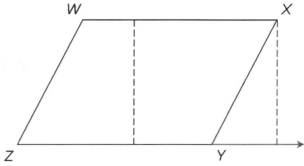

The dotted line shows height; as you can see, it's not slanted at all.

Practice Questions

1. What is the area of a parallelogram with base 100 and height 6?

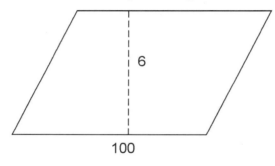

2. What is the area of a parallelogram with base 14 and height 8?

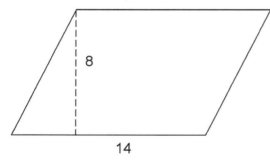

3. What is the area of the parallelogram shown?

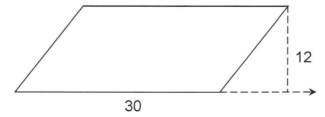

Solutions

1. To find the area, multiply base (100) $\times$ height (6).

The area equals 100 $\times$ 6, or 600 square units.

2. To find the area, multiply base $\times$ height. The area equals 14 $\times$ 8, or 112 square units.

3. The parallelogram has a base of 30 and a height of 12.

The area equals 30 $\times$ 12, or 360 square units.

Trapezoids

Trapezoids are quadrilaterals with two parallel sides. The remaining sides are not parallel.

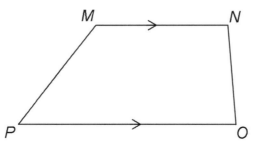

Plane Geometry

An **isosceles trapezoid** has two equal, non-parallel sides and two equal angles.

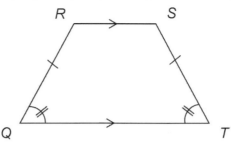

Finding perimeter

The perimeter of a trapezoid is found by adding up the lengths of each side, like with the other figures we have seen.

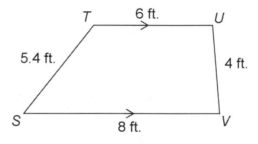

The perimeter of this trapezoid is 6 + 4 + 8 + 5.4, or 23.4 feet.

Finding area

To find the area of a trapezoid, we must first find its bases and its height.

Base and height

A trapezoid has two bases. The parallel sides are its bases.

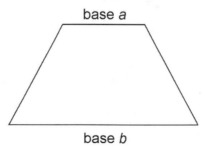

The height of the trapezoid is the distance from one base to the other.

egghead's Guide to Geometry

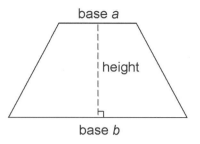

base *a*

height

base *b*

Formula

To find the area, we multiply the average of the two bases times the height:

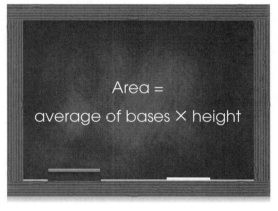

Area =
average of bases × height

Here's the shortened form:

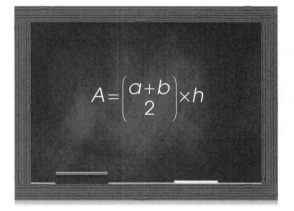

$$A = \left(\frac{a+b}{2}\right) \times h$$

Examples

The trapezoid shown has a base of 20, another base of 30, and a height of 25.

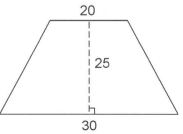

20

25

30

Plane Geometry

To find the area, we first take the average of the bases. Add together the two bases and divide by 2: 20 + 30 equals 50, and 50 divided by 2 equals 25. Multiply this number by the height: 25 times 25 equals 625 square units.

Trapezoid *JKLM* has one base of 4 meters, one base of 7 meters, and a height of 5 meters.

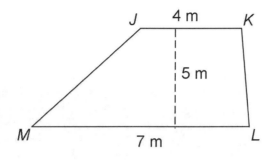

To find the area, multiply the average of the bases times the height:

$$A = \left(\frac{a+b}{2}\right) \times h$$

$$= \left(\frac{4+7}{2}\right) \times 5$$

$$= \left(\frac{11}{2}\right) \times 5$$

$$= 5.5 \times 5$$

$$= 27.5$$

The area is 27.5 square meters.

Practice Questions

1. What is the area of trapezoid *QRST* shown in the figure?

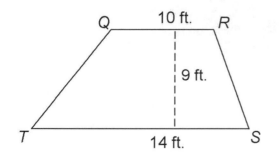

2. What is the area of the trapezoid shown?

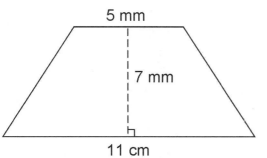

3. An isosceles trapezoid has bases that measure 14 inches and 26 inches. Its area is 400 square inches. What is the height, in inches, of the trapezoid?

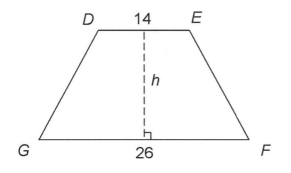

Solutions

1. The area is 108 square feet.

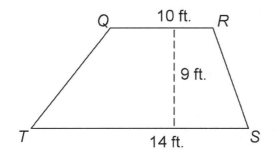

Use the formula for the area of a trapezoid:

$$A = \left(\frac{a+b}{2}\right) \times h$$

$$= \left(\frac{10+14}{2}\right) \times 9$$

$$= \left(\frac{24}{2}\right) \times 9$$

$$= 12 \times 9$$

$$= 108$$

2. The area of the trapezoid is 56 square millimeters.

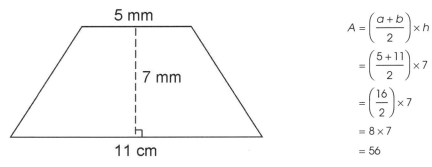

$$A = \left(\frac{a+b}{2}\right) \times h$$

$$= \left(\frac{5+11}{2}\right) \times 7$$

$$= \left(\frac{16}{2}\right) \times 7$$

$$= 8 \times 7$$

$$= 56$$

3. Insert the numbers you know into the formula to find the height of the parallelogram. The height measures 20 inches.

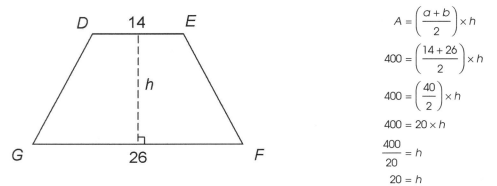

$$A = \left(\frac{a+b}{2}\right) \times h$$

$$400 = \left(\frac{14+26}{2}\right) \times h$$

$$400 = \left(\frac{40}{2}\right) \times h$$

$$400 = 20 \times h$$

$$\frac{400}{20} = h$$

$$20 = h$$

Rhombuses

The last four-sided figure we will study in depth is the rhombus. A **rhombus** is a parallelogram with four equal sides.

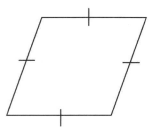

Its opposite sides are parallel, and its opposite angles are equal.

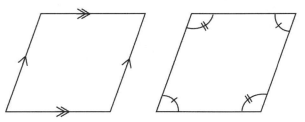

Finding perimeter

To find the perimeter of a rhombus, add up the lengths of its sides.

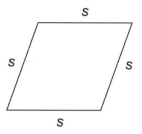

Since the sides are equal, we can also multiply the length of one side by 4. The perimeter of this rhombus is 4s.

Finding area

Formula

To find the area of a rhombus, multiply the base by the height.

Area = base × height

Base and height

The base of a rhombus is the length of its side. To find the height, draw a line perpendicular to two parallel sides.

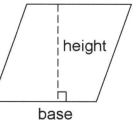

base

Examples

Rhombus *ABCD* shown has a side length of 6 feet and a height of 8 feet.

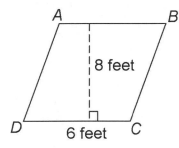

Its area equals 6 feet multiplied by 8 feet, or 48 square feet.

Practice Questions

1. What is the area of rhombus *WTUV* shown?

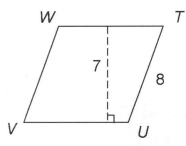

2. Find the area of rhombus *PQNO*.

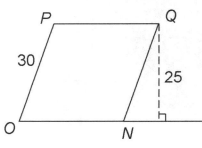

3. Rhombus *DEFG* has an area of 340 square millimeters. What is the length of its sides?

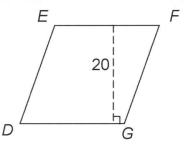

Solutions

1. The area of rhombus *WTUV* is 56 square units.

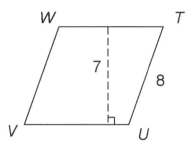

The base of a rhombus is equal to its side length, which is shown as 8 units. Multiply 8 by the height, 7 units, to obtain the area.

2. The area of the rhombus is 750 square units.

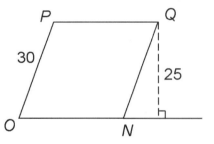

The side length is given as 30, and the height is 25. Multiply these to find the area: 30 ✕ 25 = 750.

3. Each side of the rhombus measures 17 millimeters.

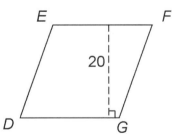

You are given the area, 340 square millimeters. Using the area formula, solve for the length of the side:

$$A = b \times h$$
$$340 = b \times 20$$
$$\frac{340}{20} = b$$
$$b = 17$$

Chapter Review

1. Which side of the rectangle is parallel to side *WX*?

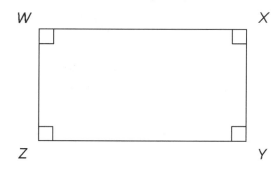

2. Find the perimeter of rectangle *KLMN*.

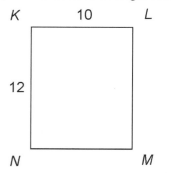

3. What is the area of rectangle *EFGH*?

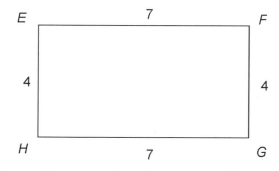

4. What is the perimeter of square *PQRS*?

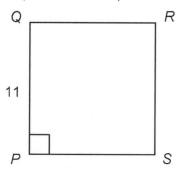

5. What is the area of a parallelogram with base 12 and height 7?

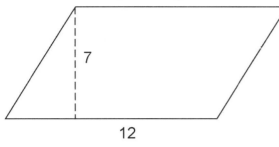

6. What is the area of the trapezoid shown?

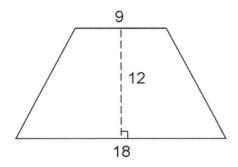

7. What is the area, in square inches, of trapezoid *SUVW* shown?

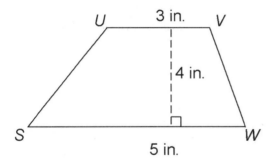

8. A building has an outdoor brick courtyard that is shaped like a rhombus. If the courtyard has the dimensions shown, what is its area?

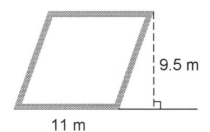

Plane Geometry

9. What is the perimeter of parallelogram *UVWX* shown in the figure?

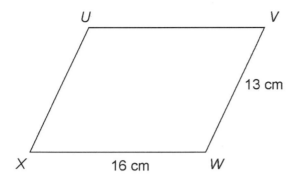

10. What is the perimeter of trapezoid *MNOP* shown?

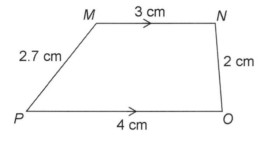

11. Rows of flowers are planted to create a garden in the shape of a rhombus, as shown in the figure. If the area of the garden is 143 square feet, what is the length of one of its sides?

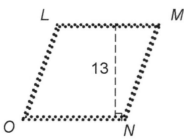

12. Kurt builds a fence around his back yard, which is in the shape of a rectangle. The fence requires 140 feet of fencing altogether. What is the width of Kurt's back yard?

50 feet

13. Maya walks around the entire perimeter of her neighbor's patio. The area of the patio is 36 square yards, and the patio has a rectangular shape, as shown. How many yards does Maya walk?

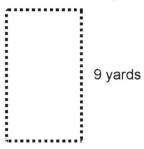

9 yards

14. A square has a side length of $2s$ centimeters, as shown in the figure. If s equals 21, what is the area of the square?

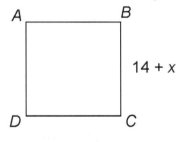

$2s$ cm

15. Quadrilateral $ABCD$, shown in the figure, is a square. What is the perimeter of $ABCD$?

A   B

$14 + x$

D   C

Solutions

1. Sides *WX* and *ZY* are parallel.

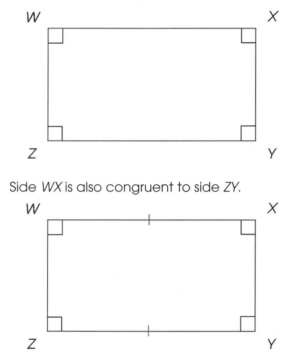

Side *WX* is also congruent to side *ZY*.

2. The opposite sides of rectangles are equal. So, side *NM* measures 10. Side *LM* measures 12.

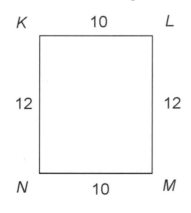

To find the perimeter, add the lengths of the sides: 10 + 12 + 10 + 12.

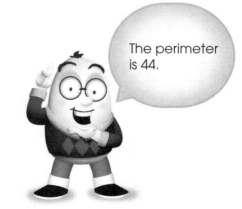

The perimeter is 44.

3. To find the area, multiply length ✕ width. The area is 28 square units.

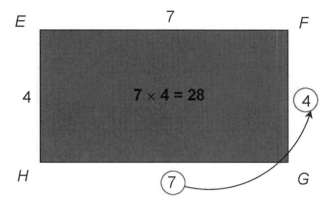

4. To find the perimeter, add together all four sides. All sides of a square are equal, so each side measures 11.

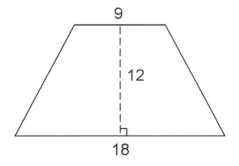

The perimeter is 11 + 11 + 11 + 11, or 44.

5. To find the area, multiply base ✕ height. The area equals 12 ✕ 7, or 84 square units.

6. The area of the trapezoid is 162 square units.

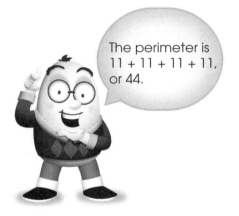

Using the formula for the area of a trapezoid, substitute 9 units and 18 units for the bases and 12 units for the height:

$$A = \left(\frac{a+b}{2} \right) \times h$$

$$= \left(\frac{9+18}{2} \right) \times 12$$

$$= \left(\frac{27}{2} \right) \times 12$$

$$= 13.5 \times 12$$

$$= 162$$

Plane Geometry

7. The area of trapezoid *SUVW* is 16 square inches.

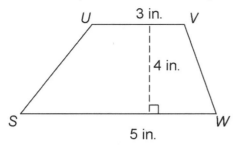

Here we would use 3 and 5 for the bases and 4 inches for the height:

$$A = \left(\frac{a+b}{2}\right) \times h$$

$$= \left(\frac{3+5}{2}\right) \times 4$$

$$= \left(\frac{8}{2}\right) \times 4$$

$$= 4 \times 4$$

$$= 16$$

8. The area of the courtyard is 104.5 square meters.

To find the area, multiply the base of the rhombus times its height. The base is equal to the length of the side, 11 meters, and the height is 9.5 meters, as shown. The area is 11 × 9.5, or 104.5 m².

9. The perimeter of parallelogram *UVWX* is 58 centimeters. Add up the lengths of the sides: 16 + 13 + 16 + 13 = 58 centimeters.

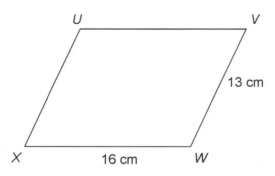

10. The perimeter of trapezoid *MNOP* is 11.7 centimeters.

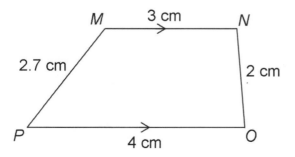

Add up the lengths of the sides: 3 + 2 + 4 + 2.7 = 11.7 centimeters.

11. One side of the garden measures 11 feet.

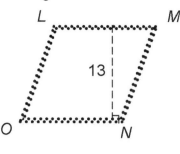

The area of the garden is 143 square feet. Using the formula for the area of a rhombus, solve for the length of the side.

$$A = b \times h$$
$$143 = b \times 13$$
$$\frac{143}{13} = b$$
$$11 = b$$

The base of the rhombus is the length of one side, so each side measures 11 feet.

12. The width of Kurt's back yard is 20 feet.

50 feet

To create the fence, Kurt must use 140 feet of fencing. This tells us the perimeter of the yard. Use the formula for the perimeter of a rectangle to find the width of the yard:

$$P = 2l + 2w$$
$$140 = 2(50) + 2w$$
$$140 = 100 + 2w$$
$$140 - 100 = 2w$$
$$40 = 2w$$
$$\frac{40}{2} = w$$
$$20 = w$$

13. Maya walks a total of 26 yards.

9 yards

The area of the patio is 36 square yards. Using the formula for the area of a rectangle, solve for the length of the patio:

$$A = l \times w$$
$$36 = l \times 9$$
$$\frac{36}{9} = l$$
$$4 = l$$

The length of the patio is 4 yards. Now, use the perimeter formula to determine the distance Maya walks:

$$P = 2l + 2w$$
$$= 2(4) + 2(9)$$
$$= 8 + 18$$
$$= 26$$

Plane Geometry

14. The area of the square is 1,764 square centimeters.

2s cm

Substitute 21 for the variable *s* to determine the length of each side. Each side measures 2 times 21, or 42 centimeters. Now, use the formula for the area of a square:

$$A = \text{side} \times \text{side}$$
$$= 42 \times (42)$$
$$= 1{,}764$$

15. The perimeter of *ABCD* is 56 + 4*x*.

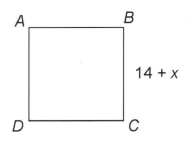

In this figure, we are given the length of the side as 14 + *x*. We don't know the value of *x*, so keep *x* as a variable when solving for the perimeter.

The perimeter of a square is equal to 4 times the length of its side. Multiply 4 times 14 + *x*:

$$P = 4 \times s$$
$$= 4(14 + x)$$
$$= (4 \times 14) + (4 \times x)$$
$$= 56 + 4x$$

egghead's Guide to Geometry

Chapter 4

Triangles

Hi! I'm egghead. I will teach the following concepts in this chapter:

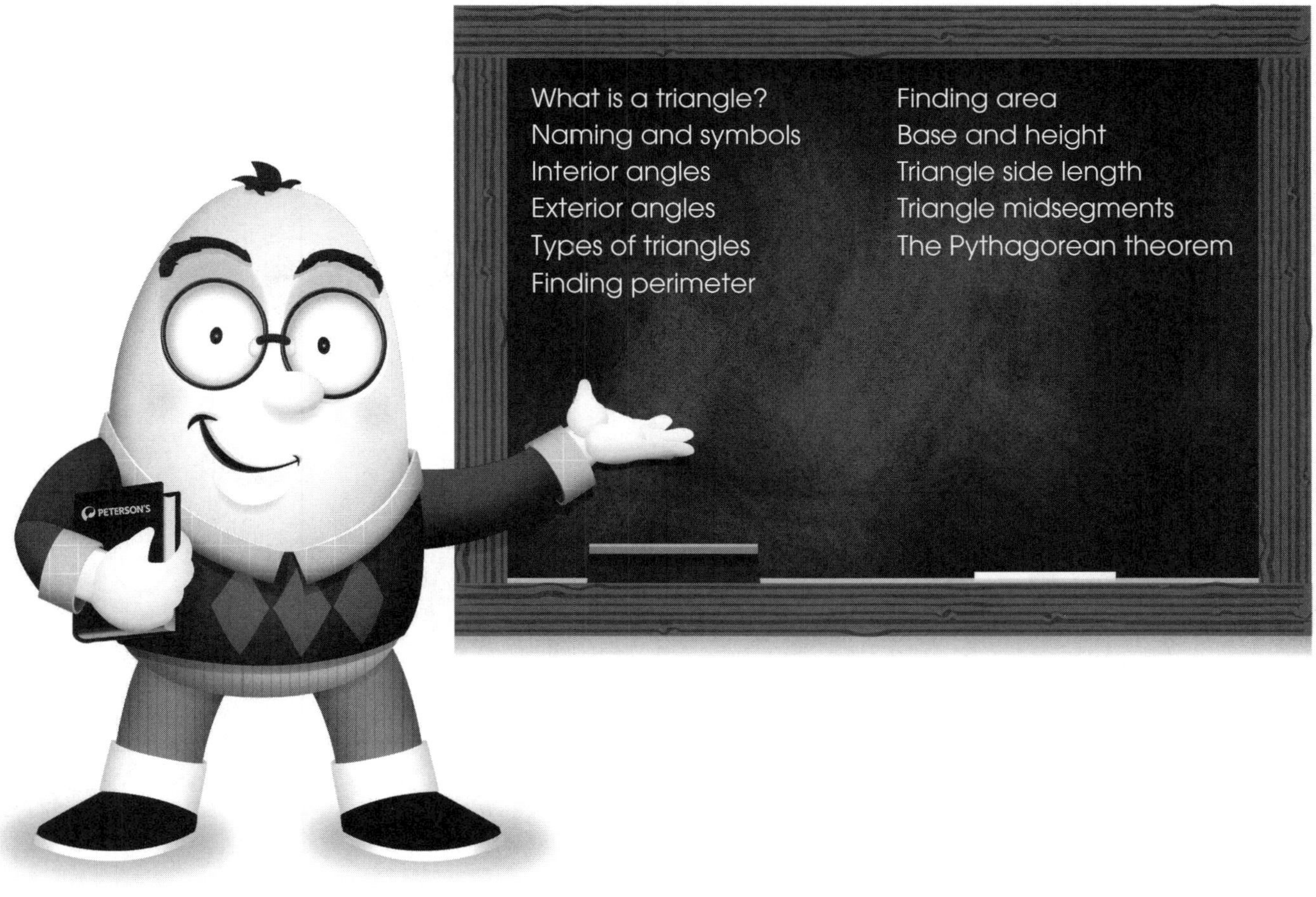
What is a triangle?
Naming and symbols
Interior angles
Exterior angles
Types of triangles
Finding perimeter
Finding area
Base and height
Triangle side length
Triangle midsegments
The Pythagorean theorem
PETERSON'S

What is a triangle?

A triangle is a shape with three sides.

The root *tri-* means "three" in Greek and Latin.

Triangles have three sides and three angles. Here are some examples:

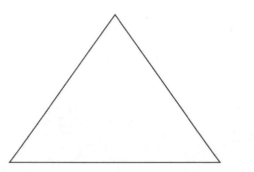
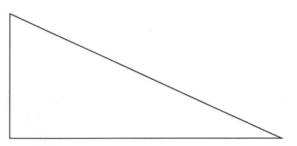

Naming and symbols

To name a triangle, use its points or **vertices.**

Triangles have three vertices. Each triangle therefore has three letters in its name. When naming a triangle in short form, use the symbol △. You can begin at any vertex.

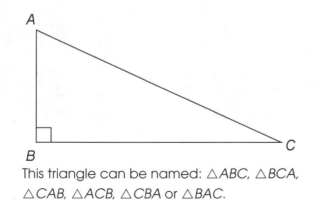

This triangle can be named: △ABC, △BCA, △CAB, △ACB, △CBA or △BAC.

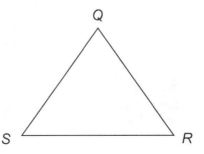

This triangle can be named: △QRS, △RSQ, △SQR, △QSR, △SRQ or △RQS.

Interior angles

There's one important property triangles have that is crucial to know:

The measures of the inside angles of a triangle add up to 180°.

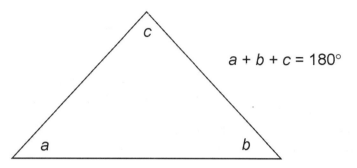

$$a + b + c = 180°$$

These inside angles are called **interior angles.**

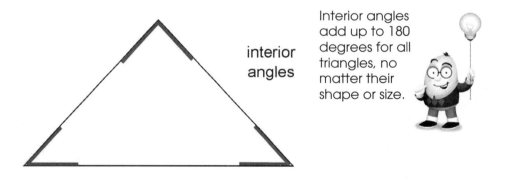

interior angles

Interior angles add up to 180 degrees for all triangles, no matter their shape or size.

Exterior angles

Triangles also have exterior angles. **Exterior angles** are formed when one side of a triangle is extended in a straight line:

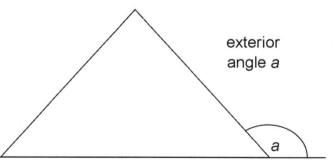

exterior angle a

Plane Geometry

Exterior angles are supplementary to their adjacent interior angles. In the triangle shown, exterior angle *a* is supplementary to interior angle *d*. Their measures add up to 180°.

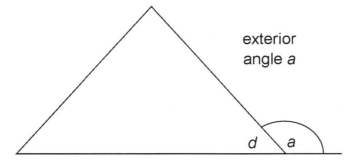

The measure of each exterior angle is also equal to the sum of the two *non-adjacent* interior angles of the triangle. In the triangle shown, m∠*a* = m∠*b* + m∠*c*.

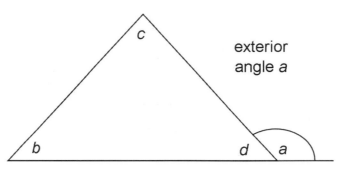

Types of triangles

Triangles can be classified by the size of their angles or lengths of their sides. Triangles classified by their sides include:

1. Isosceles triangles

2. Equilateral triangles

3. Scalene triangles

Triangles classified by their angles include:

1. Right triangles

2. Acute triangles

3. Obtuse triangles

Once you know the properties of triangles, it's easy to tell them apart.

Isosceles triangles

Isosceles triangles have two equal sides.

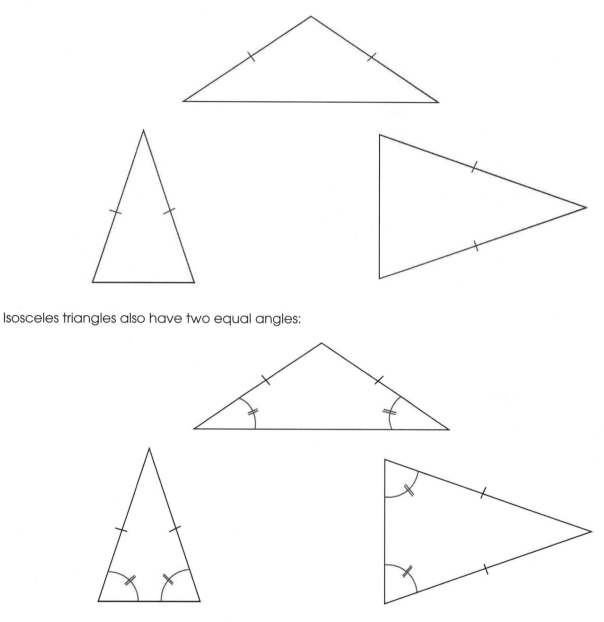

Isosceles triangles also have two equal angles:

The equal angles lie across from the equal sides!

Equilateral triangles

Equilateral triangles have three equal sides.

That's why they're called "*equi*-lateral."

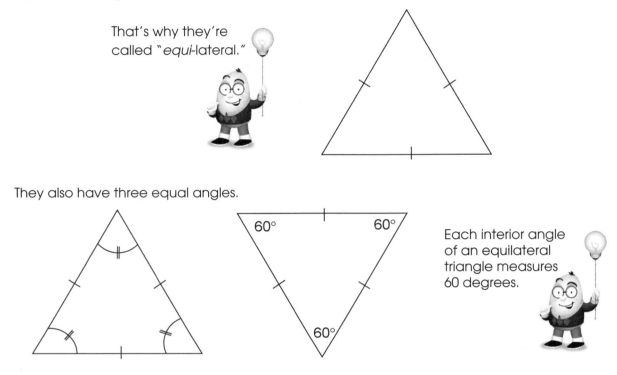

They also have three equal angles.

60°   60°

60°

Each interior angle of an equilateral triangle measures 60 degrees.

Scalene triangles

Scalene triangles are triangles that have three sides of different lengths. They have no equal angles.

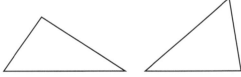

Right triangles

Right triangles have one right angle.

The side across from the right angle is the longest side of the triangle. It is called the **hypotenuse.**

Acute triangles

An acute triangle has three angles less than 90°.

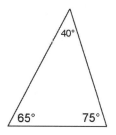

Obtuse triangles

An obtuse triangle has one of its interior angles measuring more than 90°.

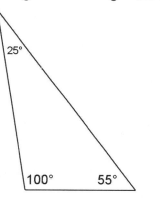

Practice Questions

1. Which one of the triangles shown is a right triangle?

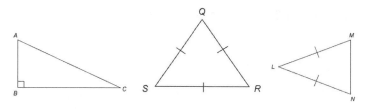

2. Which type of triangle is △*LMN*? Write the correct name in the space below the figure.

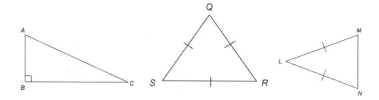

3. What is the measure of ∠V?

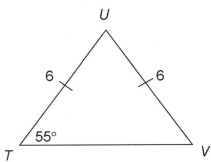

4. Triangle *MNO* is an equilateral triangle. What is the length of *MO*?

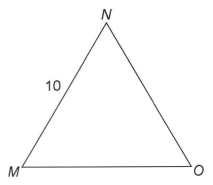

5. The triangle shown is a scalene triangle. What is the value of *x*?

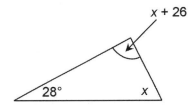

6. In the triangle shown, what is the measure of exterior angle *n*?

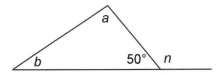

Solutions

1.

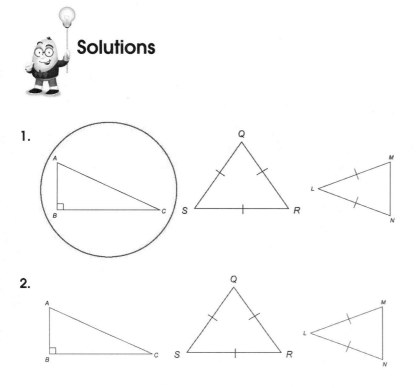

2.

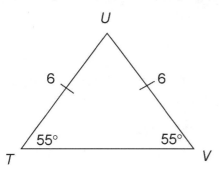

isosceles

3. This is an isosceles triangle. It has two equal sides and two equal angles. The equal angles lie across from the equal sides. Side *TU* equals side *VU,* so ∠*T* equals ∠*V.*

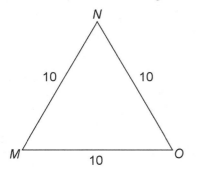

4. *MNO* is an equilateral triangle. All three sides are equal.

5. The value of *x* is 63°.

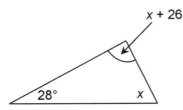

The triangle is a scalene triangle. This means no sides or angles are equal. However, we do know that the sum of the interior angles of a triangle is 180°. Set up an equation and solve for *x*:

$$28 + x + x + 26 = 180$$
$$28 + 2x + 26 = 180$$
$$54 + 2x = 180$$
$$2x = 180 - 54$$
$$2x = 126$$
$$x = 63$$

6. The measure of *n* is 130°.

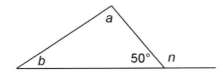

Angle *n* is an exterior angle, so it is supplementary to the angle that measures 50°. The measures of these two angles add up to 180°. To find the measure of angle *n*, use subtraction: 180° – 50° = 130°.

Finding perimeter

Now that you've learned the different types of triangles, let's take a look at perimeter questions. Perimeter questions for triangles are similar to those you would see for quadrilaterals.

The perimeter of a triangle is the distance around its outside edge. The dotted arrows show the perimeter of the triangle below:

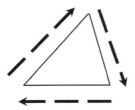

Formula

To find the perimeter, add up the lengths of the three sides:

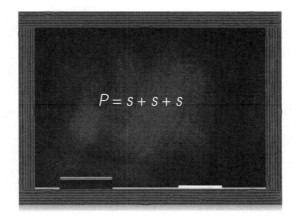

$$P = s + s + s$$

Since we've already done this quite a bit with quadrilaterals, we'll now try some more challenging questions.

Practice Questions

1. Triangle *HIJ* is isosceles. What is the perimeter of △*HIJ*?

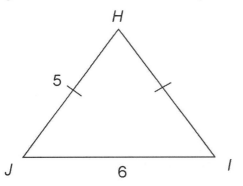

2. The figure below is an equilateral triangle. Find its perimeter.

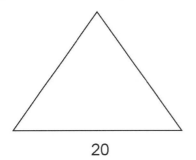

3. What is the perimeter of triangle *QRS*?

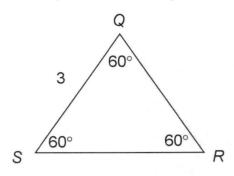

4. What is the perimeter of the triangle shown?

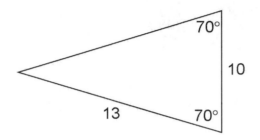

5. The triangle shown has a perimeter of 22 inches. What are the lengths of its sides?

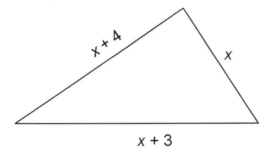

6. If the perimeter of the triangle shown is 29 centimeters, what is the value of *a*?

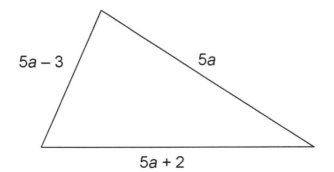

Solutions

1. An isosceles triangle has two equal sides. The figure shows that side *JH* is congruent to side *IH*. That means side *IH* measures 5. The perimeter is 5 + 5 + 6, or 16.

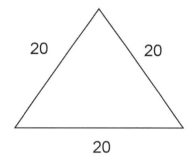

To find the perimeter, add up the three sides.

2. An equilateral triangle has three equal sides. That means all sides measure 20.

The perimeter is 20 + 20 + 20, or 60.

3. A triangle with three equal angles is an equilateral triangle. All sides are also equal.

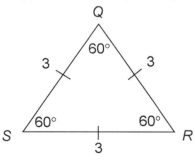

The perimeter is 3 + 3 + 3, or 9.

Plane Geometry

4. A triangle with two equal angles is an isosceles triangle. The sides opposite the equal angles are also equal. The missing side must therefore equal 13:

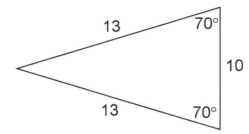

The perimeter is 13 + 13 + 10, or 36.

5. The sides of the triangle measure 5 inches, 8 inches, and 9 inches.

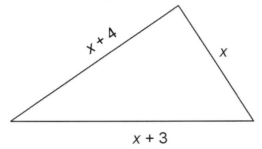

Using the perimeter formula, write an equation: $x + (x + 4) + (x + 3) = 22$. Solve for x:

$$x + (x + 4) + (x + 3) = 22$$
$$x + x + 4 + x + 3 = 22$$
$$x + x + x + 4 + 3 = 22$$
$$3x + 7 = 22$$
$$3x = 22 - 7$$
$$3x = 15$$
$$x = 5$$

The value of x is 5. So, the lengths of the sides are 5 inches, $5 + 3 = 8$ inches, and $5 + 4 = 9$ inches.

6. The value of a is 2 centimeters.

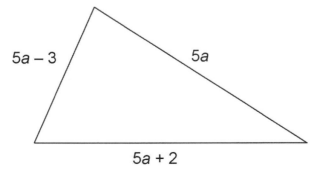

Use the perimeter formula to set up an equation: $5a + (5a - 3) + (5a + 2) = 29$. Then solve for a:

$$5a + (5a - 3) + (5a + 2) = 29$$
$$5a + 5a - 3 + 5a + 2 = 29$$
$$5a + 5a + 5a - 3 + 2 = 29$$
$$15a - 1 = 29$$
$$15a = 30$$
$$a = 2$$

We are just asked for the value of a, not the length of the sides, so we can stop here. The value of a is 2 centimeters.

Finding area

Nice work with those more challenging perimeter questions. Often, test questions in geometry require you to find a missing length, based on what you know about the shape.

In addition to asking about perimeter, triangle questions also ask about area.

The area of a triangle is the space the triangle covers. The shaded area shows the area of the triangle below:

Formula

The area of a triangle equals $\frac{1}{2}$ of its base × height.

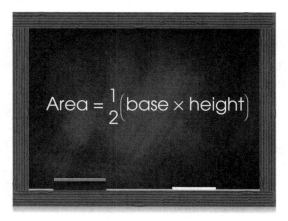

$$\text{Area} = \frac{1}{2}(\text{base} \times \text{height})$$

In shorthand, the formula looks like this:

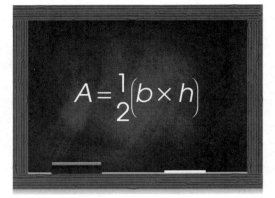

$$A = \frac{1}{2}(b \times h)$$

It can also be written like this:

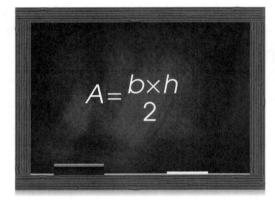

$$A = \frac{b \times h}{2}$$

When you multiply a value by $\frac{1}{2}$, it's the same as dividing that value by 2.

Base and height

The base and height of a right triangle are easiest to find. They are the sides next to the right angle:

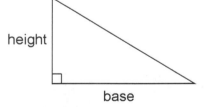

As long as you use the sides next to the right angle, and not the hypotenuse, it doesn't matter which side you call the base and which side you call the height:

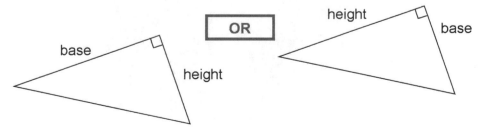

You are multiplying them, so the area still comes out the same!

egghead's Guide to Geometry

Practice Questions

1. What is the area of $\triangle ABC$?

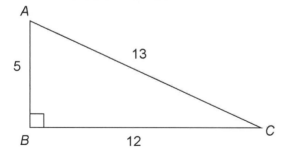

2. What is the area of $\triangle EFG$?

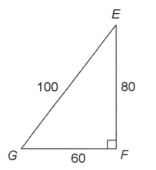

3. If $VW = 7$ inches and $XW = 7$ inches, what is the area of $\triangle VWX$?

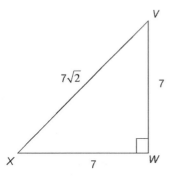

4. The triangle shown has sides measuring $2x$ feet, y feet, and $3z$ feet. What is the area of the triangle?

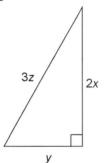

Plane Geometry

Solutions

1. The area is $\frac{1}{2}(12 \times 5)$, or 30 square units.

2. The area is $\frac{1}{2}(60 \times 80)$, or 2,400 square units.

3. The area of the triangle is 24.5 square inches.

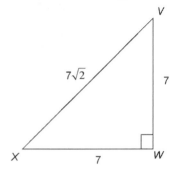

To find the area, substitute 7 for b and 7 for h into the area formula: $A = \frac{1}{2}(b \times h)$

$$= \frac{1}{2}(7 \times 7)$$
$$= \frac{49}{2}$$
$$= 24.5$$

4. The area of the triangle is xy square feet.

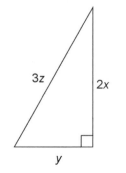

Use the area formula, $A = \frac{1}{2}(b \times h)$. For this question, you're only given variables and not actual numbers, so substitute the variables in for the base and the height. Plug in y for the base and $2x$ for the height:

$$A = \frac{1}{2}(b \times h)$$
$$= \frac{1}{2}(y \times 2x)$$
$$= \frac{1}{2}(2xy)$$
$$= \frac{2xy}{2}$$
$$= xy$$

More heights

With isosceles and equilateral triangles, finding the height is a little more complex.

Like with parallelograms, you can use one side of the triangle as the base, but you can't use another side of the triangle as its height:

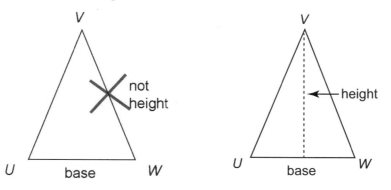

The dotted line shows the height of the triangle. Here is another example:

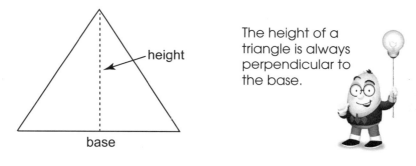

The height of a triangle is always perpendicular to the base.

Practice Questions

1. What is the area of triangle *HIJ*?

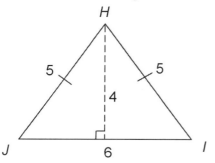

2. What is the area of △*LMN*?

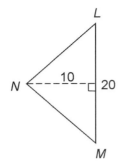

3. In △*QRS* shown, the length of $\overline{SR}$ is 13 inches. The area of the triangle is 52 square inches. What is the length of $\overline{QP}$?

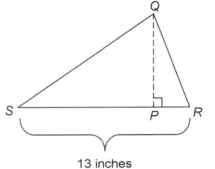

13 inches

4. The area of △*GHJ* is 47.5 square feet. What is the length of $\overline{GK}$?

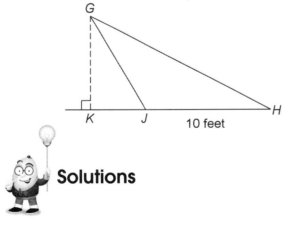

Solutions

1. The area equals $\frac{1}{2}$ of the base times the height. The dotted line shows the height, 4. The area is:

$$Area = \frac{1}{2}(base \times height)$$

$$= \frac{1}{2}(6 \times 4)$$

$$= \frac{1}{2}(24)$$

$$= 12$$

2. The base, *LM*, is 20. The height, shown by the dotted line, is 10. The area is:

$$\text{Area} = \frac{1}{2}(\text{base} \times \text{height})$$

$$= \frac{1}{2}(20 \times 10)$$

$$= \frac{1}{2}(200)$$

$$= 100$$

3. The length of $\overline{QP}$ is 8 inches.

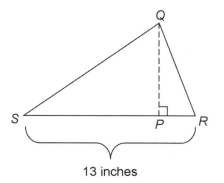

13 inches

In this triangle, $\overline{SR}$ represents the base, and $\overline{QP}$ represents the height. Using the area formula, solve for the length of $\overline{QP}$.

$$A = \frac{1}{2}(b \times h)$$

$$52 = \frac{1}{2}(13 \times h)$$

$$52 = \frac{13h}{2}$$

$$52 \times 2 = 13h$$

$$104 = 13h$$

$$h = 8$$

4. The length of $\overline{GK}$ is 9.5 feet. In this triangle, $\overline{JH}$ represents the base, and $\overline{GK}$ represents the height. Using the area formula, solve for the length of $\overline{GK}$.

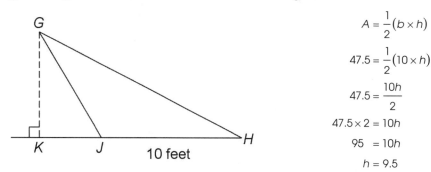

10 feet

$$A = \frac{1}{2}(b \times h)$$

$$47.5 = \frac{1}{2}(10 \times h)$$

$$47.5 = \frac{10h}{2}$$

$$47.5 \times 2 = 10h$$

$$95 = 10h$$

$$h = 9.5$$

Triangle side length

When triangles are drawn, their sides must fit a certain rule. The sum of any two sides must always be greater than the length of the remaining side.

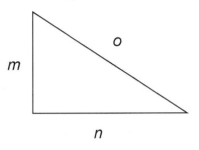

In the triangle shown, the length of *m* + *n* must be greater than the length of *o*. Similarly, *m* + *o* must be greater than *n*, and *n* + *o* must be greater than *m*.

Example

A triangle could have sides of length 6, 8, and 10. But it could not have sides of length 6, 8, and 20, because 6 + 8 is not greater than 20.

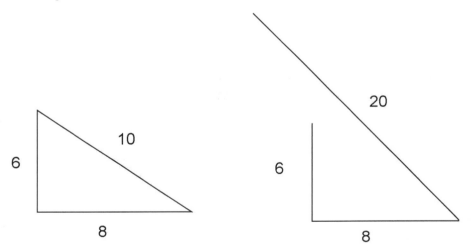

Practice Questions

1. A triangle has two sides that measure 4 centimeters and 7 centimeters. What is the largest possible whole number length of the third side?

2. A triangle has two sides that measure 3 feet and 6 feet. What is the smallest possible whole number length of the third side?

Solutions

1. The largest possible whole number length of the third side is 10 centimeters.

   The lengths of the first two legs of the triangle add up to 11 centimeters. The third leg must be smaller than 11 centimeters, so it can measure 10 centimeters at most.

2. The smallest possible whole number length of the third side is 4 feet.

   The lengths of any two sides of a triangle must always add up to more than the remaining side. The third side could not measure 3 feet, because 3 + 3 is not more than 6 feet. The third side would have to measure at least 4 feet.

Triangle midsegments

If you draw a line between the midpoints of any two sides of a triangle, you create what is known as a triangle **midsegment.** In this triangle, $\overline{DE}$ is a midsegment:

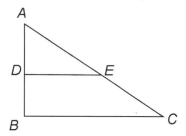

The midsegment of a triangle is always parallel to the third side and half its length.

Example

In $\triangle ABC$, midsegment *DE* is parallel to side $\overline{BC}$. Since *BC* = 10, we know that *DE* = 5.

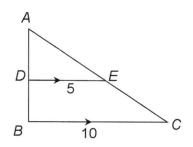

Practice Questions

1. In △STU, $\overline{UT}$ measures 9 feet. What is the length of midsegment $\overline{VW}$?

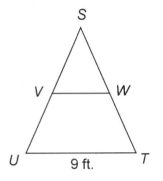

2. Midsegment $\overline{LK}$ is parallel to side $\overline{MN}$, as shown in the figure. If $\overline{ML}$ measures 8 inches, what is the length of $\overline{MO}$?

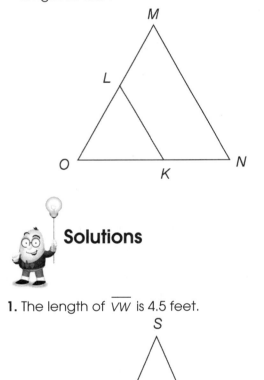

Solutions

1. The length of $\overline{VW}$ is 4.5 feet.

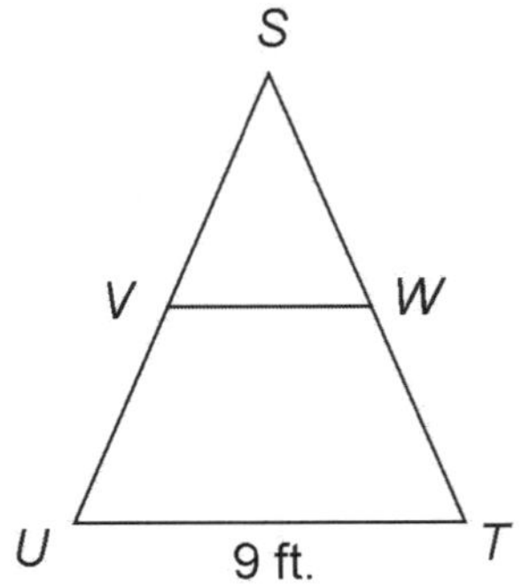

Since $\overline{VW}$ is a midsegment of the triangle, it is parallel to $\overline{UT}$ and equal to half its length. So, $\overline{VW}$ measures half of 9, or 4.5 feet.

egghead's Guide to Geometry

2. The length of $\overline{MO}$ is 16 inches.

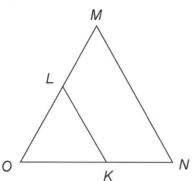

Midsegments connect the midpoints of two sides of a triangle. Thus, we know that L is a midpoint of $\overline{MO}$. If $ML = 8$ inches, then $MO = 16$ inches.

The Pythagorean theorem

Drum roll, please!

Before we leave this chapter on Triangles, there is one concept left to cover. This might be the most important concept you learn in geometry.

This concept is the **Pythagorean theorem**. It is named after the ancient Greek scholar Pythagoras, who developed it many years ago.

The Pythagorean theorem appears on geometry tests often. To do well on these tests, you must understand it. We'll cover the basics here.

Formula

Pythagoras was a brilliant scholar. He created a formula that changed math forever.

One day, Pythagoras was contemplating right triangles. He noticed an interesting relationship between the sides. He labeled the sides like this:

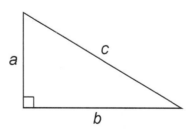

In his labeling, side c was always across from the right angle. The other two sides were sides a and b.

Pythagoras realized that if you knew the lengths of two sides of a right triangle, you could always find the length of the third side. This is how.

Plane Geometry

First, you multiply the length of side *a* by itself:

Step One

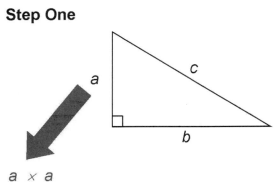

$a \times a$

This produces $a \times a$, or a^2.

Next, you multiply the length of side *b* by itself:

Step Two

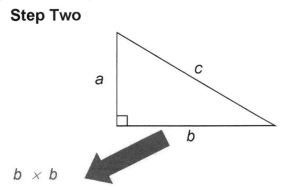

$b \times b$

This produces $b \times b$, or b^2.

Third, you multiply the length of side *c* by itself:

Step Three

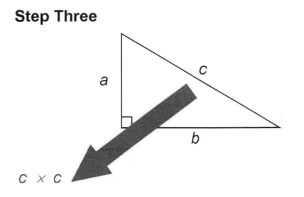

$c \times c$

This gives you c^2.

Chapter 4: Triangles

Once you have these three values, a^2, b^2, and c^2, create an equation: $a^2 + b^2 = c^2$.

Step Four

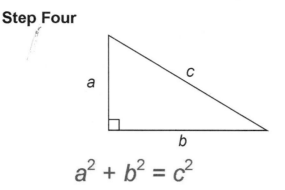

$$a^2 + b^2 = c^2$$

This is Pythagoras' famous formula. It became widely used in all of geometry.

To be honest, the "story" of how the formula was developed is . . . well, it's made up. But the theorem itself is completely true.

Example

Here is an example of how the Pythagorean theorem works. Say you want to find the length of side c in the right triangle:

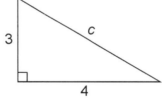

Side c is the longest side, or hypotenuse.

We know from the figure that a equals 3 and b equals 4. Plug these into the formula:

$$a^2 + b^2 = c^2$$
$$3^2 + 4^2 = c^2$$

To find the length of c, solve the equation:

$$3^2 + 4^2 = c^2$$
$$(3 \times 3) + (4 \times 4) = c^2$$
$$(9) + (16) = c^2$$
$$25 = c^2$$

Now we know that 25 equals $c \times c$. What number multiplied by itself equals 25?

$$5 \times 5 = 25$$

The length of c is 5.

Practice Questions

1. Using the Pythagorean theorem, find the length of *c* in the right triangle shown.

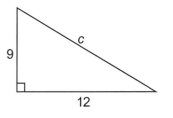

2. Using the Pythagorean theorem, find the length of *c* in the right triangle shown.

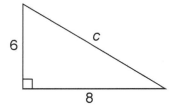

3. Using the Pythagorean theorem, find the length of *c* in the right triangle shown.

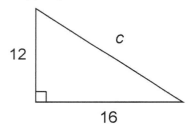

4. What is the length of the hypotenuse of the triangle shown?

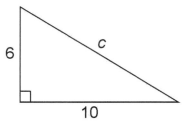

5. If $VX = 6$ and $VW = 12$, what is the length of side XW?

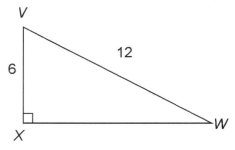

6. What is the value of *a* in the figure shown?

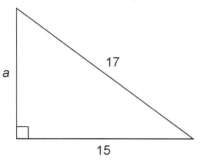

7. The triangle shown has legs measuring 20 inches and 12 inches. What is the length of the hypotenuse, *c*?

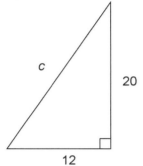

8. What is the length of $\overline{AC}$?

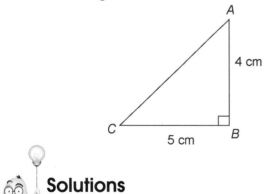

 ## Solutions

1. First, multiply the length of one side by itself. Let's start with the shorter side, 6.

Step One

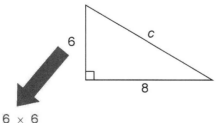

This produces 6 × 6, which can be written as 6^2.

6 × 6

Next, multiply the length of the other side by itself:

Step Two

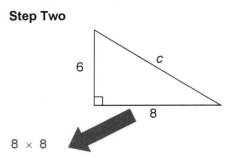

This gives us 8^2.

Third, multiply the length of side c by itself:

Step Three

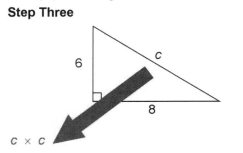

This gives us c^2.

Now, add together $6^2 + 8^2$. These are equal to c^2.

Step Four

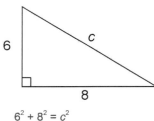

$6^2 + 8^2 = c^2$

Then solve the equation to find the length of c.

$$6^2 + 8^2 = c^2$$
$$(6 \times 6) + (8 \times 8) = c^2$$
$$(36) + (64) = c^2$$
$$100 = c^2$$

What number multiplied by itself equals 100?

$$10 \times 10 = 100$$

The length of c is 10.

2. We know from the figure that one side equals 9 and another side equals 12. Which side is *a* and which is *b*? Let's call the smaller side *a*. Plug these into the formula:

$$a^2 + b^2 = c^2$$
$$9^2 + 12^2 = c^2$$

To find the length of *c*, solve the equation:

$$9^2 + 12^2 = c^2$$
$$(9 \times 9) + (12 \times 12) = c^2$$
$$(81) + (144) = c^2$$
$$225 = c^2$$

This tells us that $c \times c$ equals 225. What number multiplied by itself equals 225?

$$15 \times 15 = 225$$

The length of *c* is 15.

3. Using the Pythagorean formula, set up an equation:

$$a^2 + b^2 = c^2$$
$$12^2 + 16^2 = c^2$$

Now, solve the equation to find the length of *c*.

$$12^2 + 16^2 = c^2$$
$$(12 \times 12) + (16 \times 16) = c^2$$
$$(144) + (256) = c^2$$
$$400 = c^2$$

What number multiplied by itself equals 400?

$$20 \times 20 = 400$$

The length of *c* is 20.

4. The length of the hypotenuse is $2\sqrt{34}$.

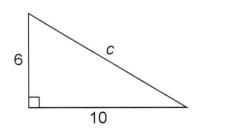

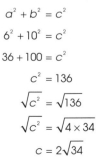

$$a^2 + b^2 = c^2$$
$$6^2 + 10^2 = c^2$$
$$36 + 100 = c^2$$
$$c^2 = 136$$
$$\sqrt{c^2} = \sqrt{136}$$
$$\sqrt{c^2} = \sqrt{4 \times 34}$$
$$c = 2\sqrt{34}$$

5. Side *XW* measures $6\sqrt{3}$ units.

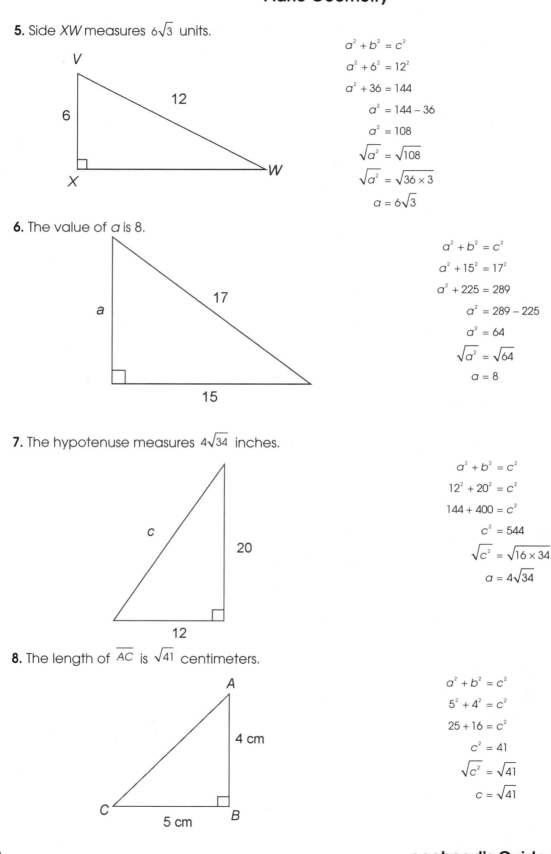

$$a^2 + b^2 = c^2$$
$$a^2 + 6^2 = 12^2$$
$$a^2 + 36 = 144$$
$$a^2 = 144 - 36$$
$$a^2 = 108$$
$$\sqrt{a^2} = \sqrt{108}$$
$$\sqrt{a^2} = \sqrt{36 \times 3}$$
$$a = 6\sqrt{3}$$

6. The value of *a* is 8.

$$a^2 + b^2 = c^2$$
$$a^2 + 15^2 = 17^2$$
$$a^2 + 225 = 289$$
$$a^2 = 289 - 225$$
$$a^2 = 64$$
$$\sqrt{a^2} = \sqrt{64}$$
$$a = 8$$

7. The hypotenuse measures $4\sqrt{34}$ inches.

$$a^2 + b^2 = c^2$$
$$12^2 + 20^2 = c^2$$
$$144 + 400 = c^2$$
$$c^2 = 544$$
$$\sqrt{c^2} = \sqrt{16 \times 34}$$
$$a = 4\sqrt{34}$$

8. The length of $\overline{AC}$ is $\sqrt{41}$ centimeters.

$$a^2 + b^2 = c^2$$
$$5^2 + 4^2 = c^2$$
$$25 + 16 = c^2$$
$$c^2 = 41$$
$$\sqrt{c^2} = \sqrt{41}$$
$$c = \sqrt{41}$$

Chapter Review

1. Triangle *CDE* is equilateral. What is the measure of ∠*D*?

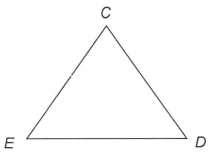

2. What is the measure of ∠*T*?

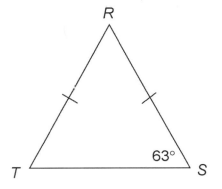

3. What is the perimeter of the triangle below?

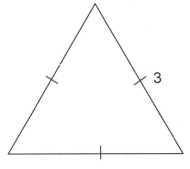

4. What is the area of the triangle shown?

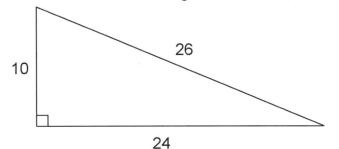

Plane Geometry

5. What is the formula for the Pythagorean theorem?

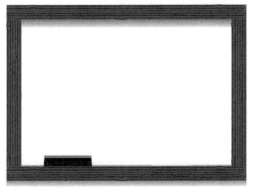

6. Three points are located in a triangular pattern, as shown in the figure. If the perimeter of the triangle is 64 kilometers, what is the distance from point *V* to point *W*?

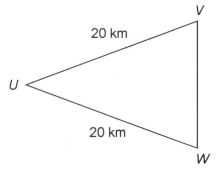

7. Right triangle *EFG* has legs measuring 9 centimeters and 7 centimeters, as shown. What is the area of the triangle?

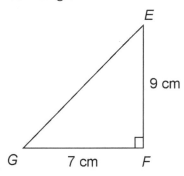

8. If ∠*H* measures 7*x* and ∠*I* measures 7*x* – 8, what is the value of *x* in the figure shown?

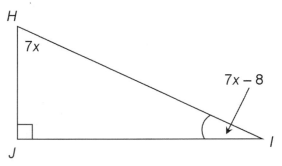

9. The triangle shown has three angles measuring 2*a*, 2*a*, and 2*a* + 18, as shown. What are the measures, in degrees, of the three angles?

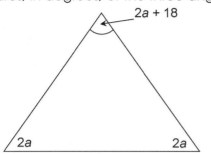

10. In the triangle shown, if *y* measures 62° and *z* measures 58°, what is the measure of exterior angle *x*?

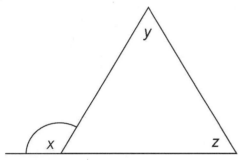

11. Allejandra cuts a wooden shelf in the shape of a triangle, as shown. The perimeter of the shelf is 37.5 inches. What are the measures, in degrees, of each of the shelf's three angles?

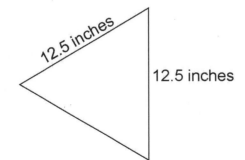

12. In the city of East Harbor, three buildings are located in a triangular arrangement, as shown. If Building *N* is 6 kilometers northwest of Building *O* and 3.5 kilometers north of Building *M*, Building *M* is approximately how many kilometers west of Building *O*?

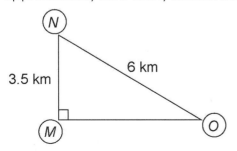

Plane Geometry

13. Victor creates a triangle by placing three metal rods in the pattern shown. The perimeter of the triangle is 16 feet. What is the length of each metal rod Victor uses?

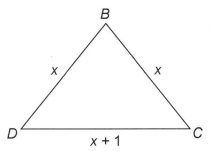

14. Jennifer cuts a piece of fabric in the shape of a triangle. The measurements are shown in the figure. What is the area of the triangle she cuts?

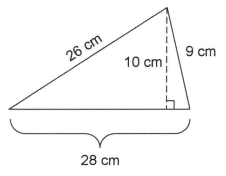

15. On the city map shown, the bookstore lies exactly halfway between the bakery and the theater. If Al's Diner is located 2 miles from both the bakery and the theater, how far is the bookstore from the bakery?

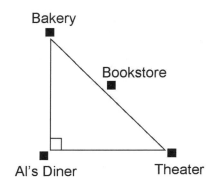

Solutions

1. Triangle *CDE* is an equilateral triangle. All of its angles measure 60.°

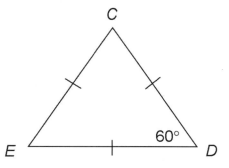

2. This is an isosceles triangle, with two equal sides and two equal angles. Side *TR* equals side *SR*, so ∠*T* equals ∠*S*. The measure of ∠*T* is 63°.

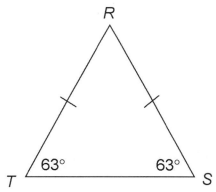

3. This triangle has three equal sides. That means all sides measure 3. The perimeter is 3 + 3 + 3, or 9.

4. To find the area, use the area formula: $\frac{1}{2}(\text{base} \times \text{height})$. The area is $\frac{1}{2}(24 \times 10)$, or 120.

5.

$$a^2 + b^2 = c^2$$

6. The distance from point *V* to point *W* is 24 kilometers.

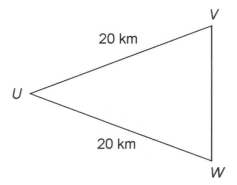

We are given the perimeter, 64 kilometers, and the lengths of two sides. Subtract the lengths of the two given sides to determine the distance from *V* to *W*: 64 kilometers – 20 kilometers – 20 kilometers = 24 kilometers.

7. The area of the triangle is 31.5 square centimeters.

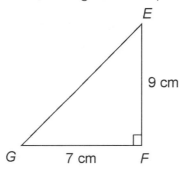

Use the area formula, $A = \frac{1}{2}(b \times h)$. Substitute 7 for the base and 9 for the height of the triangle:

$$A = \frac{1}{2}(b \times h)$$

$$= \frac{1}{2}(7 \times 9)$$

$$= \frac{1}{2}(63)$$

$$= 31.5$$

8. The value of *x* is 7°.

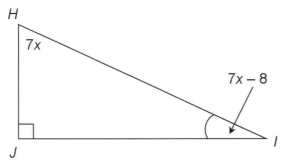

Angles *H* and *I* are complementary, so their measures add up to 90°. The remaining angle of the triangle is a right angle, and all three angles of the triangle add up to 180°, so m∠*H* plus m∠*I* must equal 90°:

$$m\angle H + m\angle I = 90$$
$$7x + (7x - 8) = 90$$
$$14x - 8 = 90$$
$$14x = 90 + 8$$
$$14x = 98$$
$$x = 7$$

9. The three angles measure 54°, 54°, and 72°.

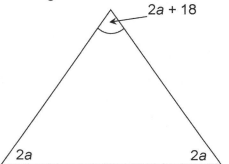

First, find the value of *a*. The measures of the interior angles of a triangle add up to 180°. Use this information to set up an equation and solve for *a*:

$$2a + 2a + 2a + 18 = 180$$
$$6a + 18 = 180$$
$$6a = 180 - 18$$
$$6a = 162$$
$$a = 27$$

Now that we know the value of *a*, we can plug this into the expressions shown. The two smaller angles measure 2 times 27, or 54°. The larger angle, at the top of the triangle, measures 2(27) + 18, or 72°.

10. The measure of exterior angle *x* is 120°.

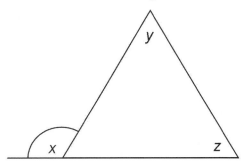

Angle *x* is an exterior angle, so its measure equals the sum of the two non-adjacent interior angles. Set up an equation to solve for *x*:

$$m\angle x = m\angle y + m\angle z$$
$$= 62 + 58$$
$$= 120$$

Plane Geometry

11. The shelf's angles each measure 60°.

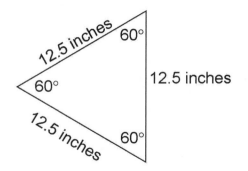

The perimeter of the shelf is 37.5 inches. Subtract the lengths of the given sides from the perimeter: 37.5 – 12.5 – 12.5 = 12.5. The remaining side of the shelf also measures 12.5 inches.

Since all three side lengths are equal, this shelf forms an equilateral triangle. The measures of its three angles are equal. Each angle therefore measures 60°.

12. Building *M* is approximately 4.873 kilometers west of Building *O*.

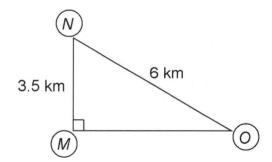

Use the Pythagorean theorem to solve this problem. Let $\overline{NM} = b$ and $\overline{NO} = c$. Solve for the measure of $\overline{MO}$, which is represented by a in the equation:

$$a^2 + b^2 = c^2$$
$$\left(\overline{MO}\right)^2 + \left(\overline{NM}\right)^2 = \left(\overline{NO}\right)^2$$
$$\left(\overline{MO}\right)^2 + (3.5)^2 = (6)^2$$
$$\left(\overline{MO}\right)^2 + 12.25 = 36$$
$$\left(\overline{MO}\right)^2 = 36 - 12.25$$
$$\left(\overline{MO}\right)^2 = 23.75$$
$$\sqrt{\left(\overline{MO}\right)^2} = \sqrt{23.75}$$
$$\overline{MO} \approx 4.873$$

egghead's Guide to Geometry

13. The lengths of the metal rods are 5 feet, 5 feet, and 6 feet.

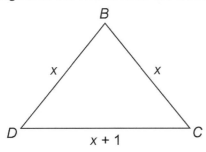

Using the formula for the perimeter of a triangle, set up an equation:

$$P = x + x + x + 1$$
$$16 = x + x + x + 1$$

Solve for x:

$$16 = x + x + x + 1$$
$$16 = 3x + 1$$
$$16 - 1 = 3x$$
$$15 = 3x$$
$$x = 5$$

The value of x is 5, so sides $\overline{BD}$ and $\overline{BC}$ each measure 5 feet. Side $\overline{CD}$ measures 5 + 1, or 6 feet.

14. The area of the fabric is 140 square centimeters.

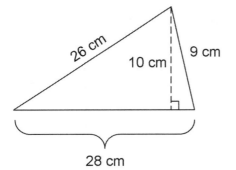

Use the length of the perpendicular line, 10 centimeters, for the height of the triangle. Use 28 centimeters as the length of the base.

$$A = \frac{1}{2}(b \times h)$$
$$= \frac{1}{2}(28 \times 10)$$
$$= \frac{1}{2}(280)$$
$$= 140$$

15. The bookstore is $\sqrt{2}$ miles from the bakery, or approximately 1.414 miles.

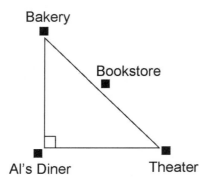

Bakery

Bookstore

Al's Diner   Theater

Use the information you are given to find the distance from the bakery to the theater. This is the length of the hypotenuse of the triangle. Plug 2 in for a and b in the Pythagorean theorem:

$$a^2 + b^2 = c^2$$
$$2^2 + 2^2 = c^2$$
$$4 + 4 = c^2$$
$$8 = c^2$$
$$\sqrt{8} = \sqrt{c^2}$$
$$\sqrt{4 \times 2} = \sqrt{c^2}$$
$$\sqrt{2 \times 2 \times 2} = \sqrt{c^2}$$
$$2\sqrt{2} = c$$

The distance from the bakery to the theater is $2\sqrt{2}$ miles. Half of this distance would be $\sqrt{2}$ miles.

Chapter 5

Similar and Congruent Triangles

Hi! I'm egghead. I will teach the following concepts in this chapter:

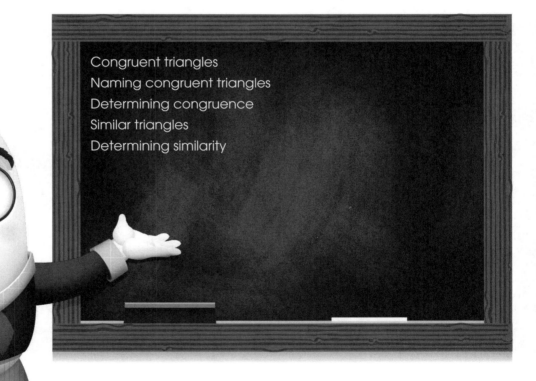

Congruent triangles
Naming congruent triangles
Determining congruence
Similar triangles
Determining similarity

Congruent triangles

Congruent triangles are triangles with three equal sides and three equal angles. When we say triangles are congruent, we're just saying the two triangles have the same size and shape.

These two triangles are congruent:

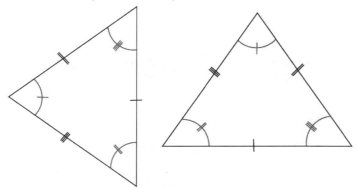

So are these two:

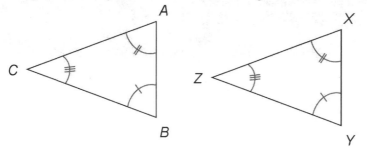

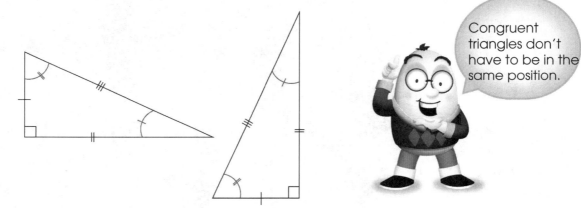

Congruent triangles don't have to be in the same position.

Corresponding sides and angles

All congruent triangles have **corresponding sides** and **corresponding angles.**

In the figure shown, ∠A corresponds to ∠X. Angle B corresponds to ∠Y, and ∠C corresponds to ∠Z. These angles are congruent.

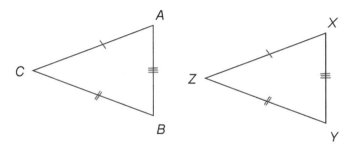

In the figure shown, side *AC* corresponds to side *XZ*. Side *CB* corresponds to side *ZY*, and side *BA* corresponds to side *YX*. These sides are congruent.

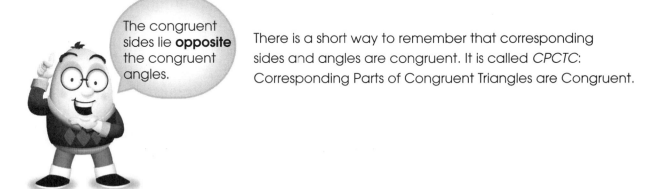

The congruent sides lie **opposite** the congruent angles.

There is a short way to remember that corresponding sides and angles are congruent. It is called *CPCTC*: Corresponding Parts of Congruent Triangles are Congruent.

Naming congruent triangles

When we name congruent triangles, it's important to make sure the vertices are listed in the correct order. The vertices of one triangle must match up with the vertices of the other triangle.

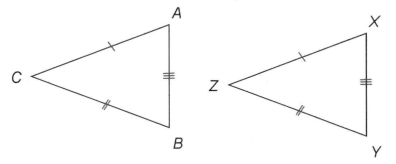

For the figure above, we could call one triangle *XYZ* and the other *XYZ*. This gives the vertices in matching order.

It would also be correct to say that △*BCA* is congruent to △*YZX*.

Determining congruence

To determine whether triangles are congruent, there are five tests that we can use. If any one of these tests applies to our triangles, we know they're congruent. We don't have to measure all six sides and all six angles.

Side-side-side

The first test is the **side-side-side** test, or SSS. If all three sides of the triangles are congruent, then the triangles are congruent.

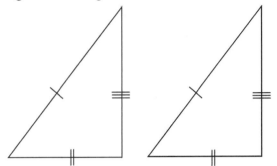

These two triangles have three equal sides. So, the triangles are congruent.

Side-angle-side

The second test is the **side-angle-side** test, or SAS. Triangles are congruent if they have two congruent sides, plus a congruent angle that lies between those sides. This angle is called an **included angle.** It is formed by the two congruent sides.

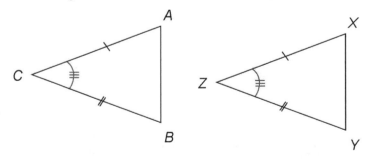

Triangles *ABC* and *XYZ* have two sets of congruent sides: $\overline{AC} \cong \overline{XZ}$ and $\overline{BC} \cong \overline{YZ}$. In addition, they also have one congruent included angle: $\angle C \cong \angle Z$. The congruent angles lie between the two congruent sides. This tells us the triangles are congruent.

Angle-side-angle

The third test is the **angle-side-angle** test, or ASA. Triangles are congruent if they have two congruent angles, plus a congruent side that lies between those angles. The congruent side is called an **included side.**

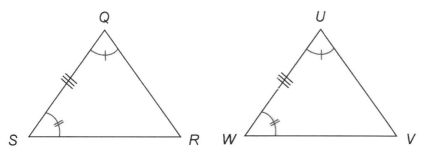

In this figure, $\angle Q \cong \angle U$ and $\angle S \cong \angle W$. We know also that sides QS and UW are congruent. Sides QS and UW lie between the congruent angles, so the triangles are congruent.

Angle-angle-side

A fourth test to prove congruence is the **angle-angle-side** test, or AAS. In this test, we examine the triangles to see if there are two congruent angles plus a congruent side that lies *across* from one of the congruent angles. If so, the triangles are congruent.

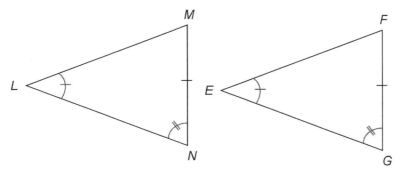

In the figure shown, angles L and E are congruent. So are angles N and G. In addition, the side across from $\angle L$ is congruent to the side across from $\angle E$. Therefore, the triangles are congruent.

Hypotenuse-leg

There is one final test to use to determine whether triangles are congruent. This is the hypotenuse-leg test, or HL.

Plane Geometry

This test is used only with right triangles.

Right triangles are congruent if they have one congruent leg and congruent hypotenuses. In the figure below, the triangles have equal hypotenuses. They also have one congruent leg, the longer leg. These triangles are therefore congruent.

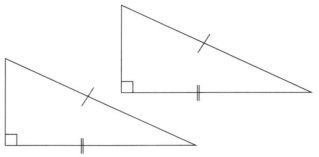

Combinations that don't work

Any of these five tests will help to determine if two triangles are congruent. However, there are some combinations that do not show congruence. These are angle-angle-angle (AAA) and side-side-angle (SSA).

If two triangles have three congruent angles, we cannot conclude that they are congruent.

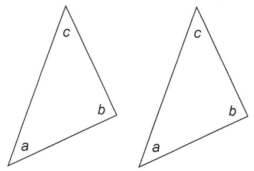

These triangles might be congruent, but we don't know for sure. The three equal angles are not enough to go on.

Similarly, if triangles have two congruent sides plus one non-included angle, this does not indicate congruence. The triangles may be congruent, but we can't know for sure that they are.

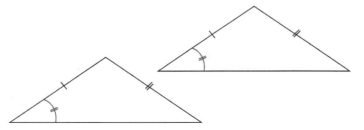

These triangles would definitely be congruent if the equal angles fell between the two equal sides.

egghead's Guide to Geometry

Examples

We know that these two triangles are congruent, because the figure shows that they have three equal sides.

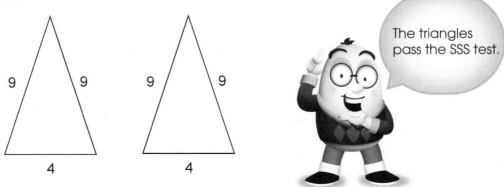

The triangles pass the SSS test.

These triangles are also congruent. They pass the ASA test:

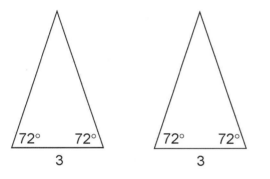

They have two congruent angles and one congruent included side.

Practice Questions

1. The triangles below are congruent. Which test proves this is so?

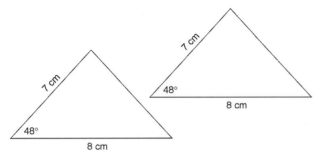

Plane Geometry

2. The triangles below are congruent. Which test proves this is so?

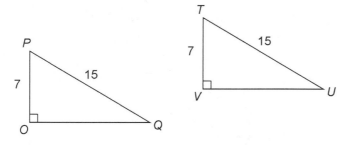

3. Are these triangles congruent? If so, list the test that proves it.

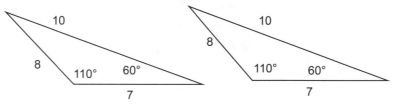

4. Which triangle is not congruent with the others?

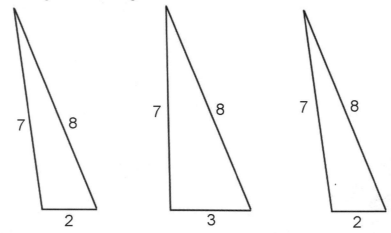

5. Which triangle is not congruent with the others?

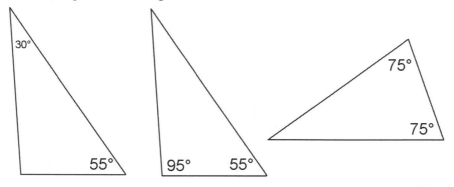

Note: Figure not drawn to scale.

Solutions

1. The triangles are congruent based on SAS, the side-angle-side test. They have two congruent sides with one congruent included angle.

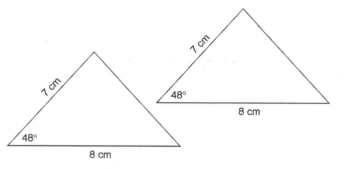

2. The triangles are congruent based on HL, the hypotenuse-leg test. They have one congruent leg and congruent hypotenuses.

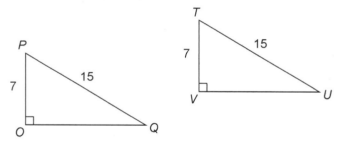

3. The triangles are congruent because they have three sets of equal sides. The side-side-side (SSS) test proves these are congruent.

4. The middle triangle is not congruent with the others, because its shortest leg measures 3 units.

5. The triangle on the right is not congruent with the other two. The left and middle triangles have angles that measure 30, 55, and 95 degrees. The triangle on the right is an isosceles triangle, with angles measuring 75, 75, and 30 degrees.

Similar triangles

Some triangles are not congruent, but instead they are similar. Similar triangles are those that are almost congruent, but not quite.

Similar triangles have three equal angles, but their sides are not equal. Instead, their sides are proportional to each other.

Ratio and proportion

To understand similar triangles, let's review the concepts of **ratio** and **proportion.**

A **ratio** is one number compared to another. To determine the ratio of two numbers, we divide the first number by the second:

The ratio of 4 to 2 is 2.

We know this because 4 divided by 2 equals 2.

Ratios are often expressed as fractions. To express the ratio 4 to 2, we write $\frac{4}{2}$.

A **proportion** is an equation between two or more ratios. To create a proportion, two ratios must be equal to each other:

$$\frac{4}{2} = \frac{8}{4}$$

In this case, both $\frac{4}{2}$ and $\frac{8}{4}$ are equal to 2.

Proportions and similar triangles

Proportions help identify similar triangles. To be similar, two triangles must have three congruent angles and three sets of proportional sides.

The ratios of the corresponding sides are equal.

In the figure below, the triangles are not congruent. One is smaller than the other.

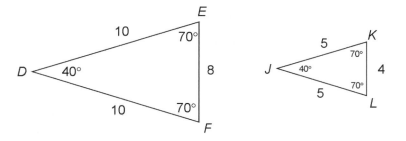

The triangles have three equal angles, and their corresponding sides are proportional. Each side of the larger triangle is twice the size of the corresponding side of the smaller triangle.

We can express the ratios of the sides in a proportion, like this:

$$\frac{DE}{JK} = \frac{EF}{KL} = \frac{FD}{LJ}$$

Substituting numbers for the side names, we see that the ratios are equal:

Each of the ratios of the corresponding sides equals 2.

Naming similar triangles

To name similar triangles, we use the corresponding sides:

$$\Delta DEF \text{ is similar to } \Delta JKL$$

In symbolic notation, we would write:

$$\Delta DEF \sim \Delta JKL$$

The symbol for similarity is ~.

Determining similarity

Two triangles are similar if three tests apply.

Angle-angle

The first test for similarity is the **angle-angle** test, or AA. If any two sets of corresponding angles of two triangles are congruent, the triangles are similar.

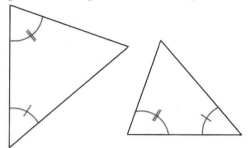

Corresponding sides are always across from equal angles.

Side-angle-side

The second test for similarity is the **side-angle-side** test, or SAS. Two triangles are similar if any two sets of corresponding sides are equally proportional and the angles between these sides are congruent.

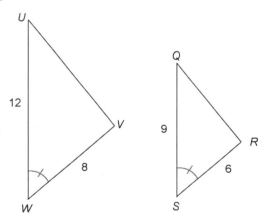

In this case, $\angle W$ is congruent to $\angle S$. Sides $\overline{QS}$ and $\overline{UW}$ are in a ratio of $\dfrac{9}{12}$, which reduces to $\dfrac{3}{4}$. Sides $\overline{RS}$ and $\overline{VW}$ are in a ratio of $\dfrac{6}{8}$, which also reduces to $\dfrac{3}{4}$. The two sets of corresponding sides are therefore equally proportional, and the triangles are similar.

Side-side-side

The third test for similarity is the **side-side-side** test, or SSS. If all three sets of corresponding sides of two triangles are proportional in the same ratio, the triangles are similar.

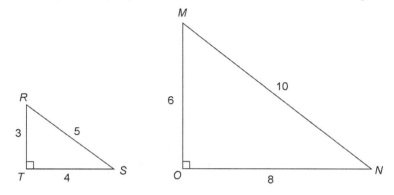

In the triangle shown, sides $\overline{MO}$ and $\overline{RT}$ are in a ratio of $\frac{6}{3}$, which reduces to 2. Sides $\overline{NO}$ and $\overline{ST}$ are in a ratio of $\frac{8}{4}$, which also reduces to 2. Sides $\overline{MN}$ and $\overline{RS}$ are in a ratio of $\frac{10}{5}$, which reduces to 2 as well. All three sets of corresponding sides have ratios in the same proportion, so the triangles are similar.

Examples

We can use the properties of similar triangles to help us find missing lengths and angle measures.

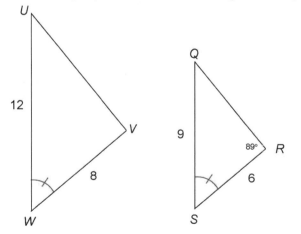

In the figure shown, m $\angle R$ = 89. Since the two triangles are similar, all of their angles are equal. So, the measure of corresponding $\angle V$ must be 89° also.

Plane Geometry

In the figure below, the two right triangles are similar.

The side of length 20 corresponds with the side of length 10, and the side of length 12 corresponds with the side of length 6.

The given sides are in a ratio of $\frac{20}{10}$ and $\frac{12}{6}$. Both of these fractions reduce to 2.

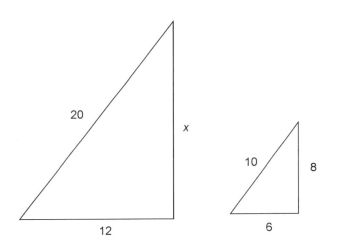

To find the length of the missing side, x, we can set up a proportion:

$$\frac{20}{10} = \frac{x}{8}$$

Cross-multiply to solve for x:

$$20 \times 8 = 10 \times x$$
$$160 = 10x$$
$$x = 16$$

The length of x is 16.

Practice Questions

1. The two triangles in the figure shown are similar. The side of length 16 corresponds with the side of length 4, and the side of length 12 corresponds with the side of length *s*. What is the length of *s*?

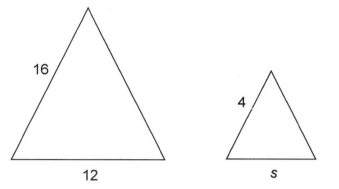

2. In the figure shown, $\triangle XYZ \sim \triangle ABC$. Find the measurement of $\angle A$.

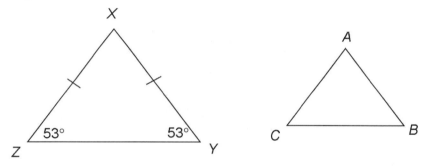

3. The two right triangles in the figure shown are similar. The side of length *y* corresponds with the side of length 7.5. What is the length of *y*?

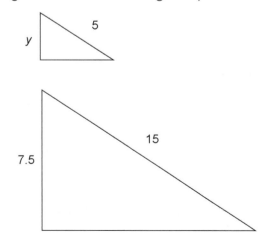

4. The two triangles in the figure have three sets of congruent corresponding angles, as shown. Find the length of *b*.

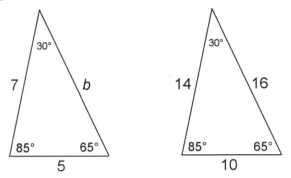

Note: Figure not drawn to scale.

5. What is the value of *c* in the figure shown?

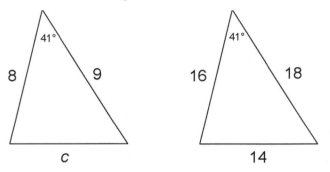

Note: Figure not drawn to scale.

 **Solutions**

1. The length of *s* is 3.

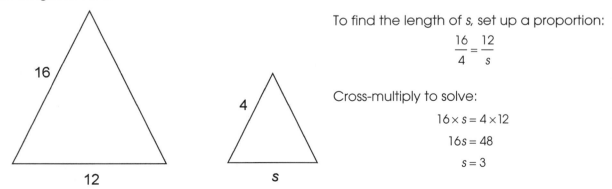

To find the length of *s*, set up a proportion:

$$\frac{16}{4} = \frac{12}{s}$$

Cross-multiply to solve:

$$16 \times s = 4 \times 12$$
$$16s = 48$$
$$s = 3$$

2. The measurement of ∠A is 74°.

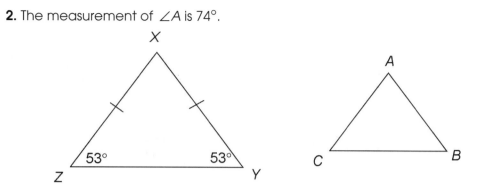

The question states that $\triangle XYZ \sim \triangle ABC$. The vertices are given in the order of correspondence. So, vertex X corresponds with vertex A. That means these two angles are equal.

To find the measurement of ∠X, subtract 53° and 53° from 180°:

$$m\angle X = 180 - 53 - 53$$
$$= 74$$

Since ∠X equals ∠A, the measure of ∠A is 74°.

3. The length of y is 2.5.

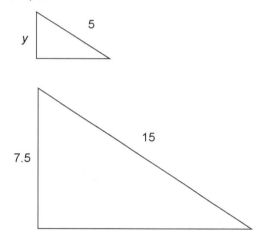

To find the length of y, set up a proportion:

$$\frac{15}{5} = \frac{7.5}{y}$$

Cross-multiply to solve:

$$15 \times y = 5 \times 7.5$$
$$15y = 37.5$$
$$y = 2.5$$

4. The side marked *b* measures 8 units.

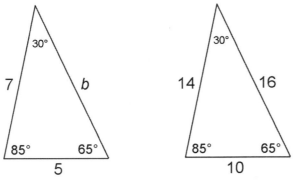

Note: Figure not drawn to scale.

The two triangles have three sets of equal angles, so they are similar triangles. Use a proportion to determine the length of *b*:

$$\frac{7}{14} = \frac{b}{16}$$
$$14 \times b = 7 \times 16$$
$$14b = 112$$
$$b = \frac{112}{14}$$
$$b = 8$$

5. The value of *c* is 7.

The two triangles are similar. We know this because they pass the side-angle-side (SAS) test. First, they have two sets of corresponding sides that are equally proportional. The triangle on the left has sides of length 8 and 9, and the triangle on the right has sides that are twice as long (16 and 18). Second, they have one set of included congruent angles that each measure 41°.

Since we know the triangles are similar, we can set up a proportion to solve for the value of *c*:

$$\frac{8}{16} = \frac{c}{14}$$
$$16 \times c = 8 \times 14$$
$$16c = 112$$
$$c = \frac{112}{16}$$
$$c = 7$$

Chapter Review

1. The triangles below are congruent. Which test proves this is so?

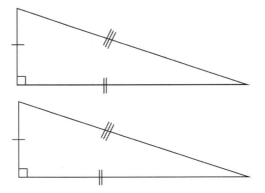

2. The triangles below are similar. Which test proves this is so?

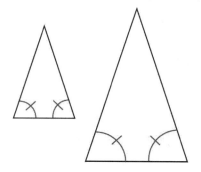

3. Are the triangles below congruent? Why or why not?

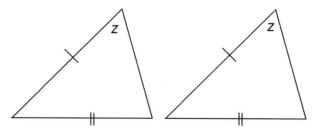

4. Are the triangles below similar? Why or why not?

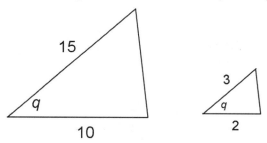

5. In the figure shown, $\triangle NMO \sim \triangle PQR$. What is the measurement of c?

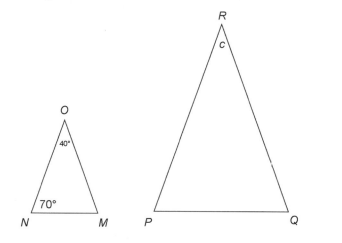

6. The two triangles in the figure shown are congruent. What is the measure of ∠R?

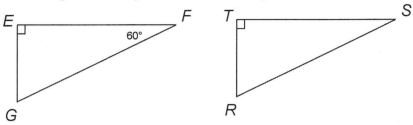

7. In the figure below, △JKL ≅ △ZYX. What is the length of *s*?

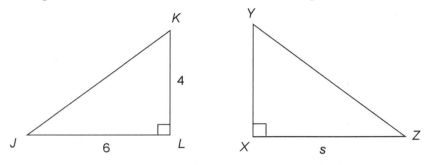

8. In the figure below, △HIJ ~ △ECD. What is the measure of ∠J?

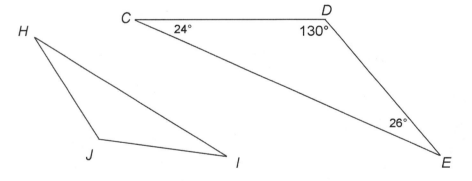

9. What is the length of $\overline{AB}$ in the figure shown?

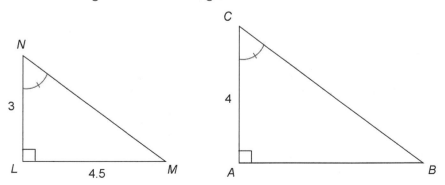

10. Find the measurement of ∠E in the figure shown.

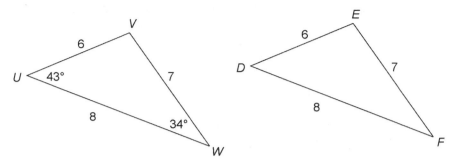

11. What is the length of $\overline{XY}$ in the figure shown?

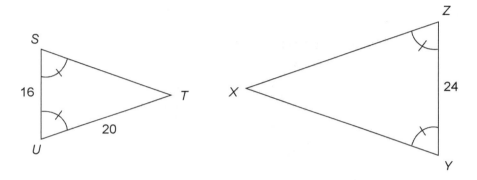

12. A college campus has buildings A, B, and C, located as shown in the figure. A sidewalk connecting the three buildings forms a triangular pattern. Buildings D, E, and F are also connected by sidewalks in a triangular pattern. The triangle formed between buildings A, B, and C is similar to the triangle formed between buildings D, E, and F. The distances are given in yards. What is the distance, in yards, between buildings B and C?

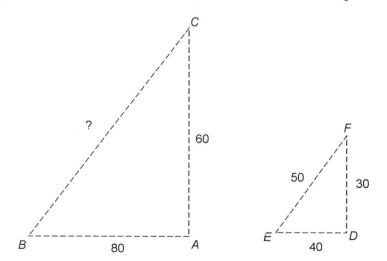

Plane Geometry

13. Carla is building a bookshelf to fit in the corner of her room. The bookshelf has two identical corner shelves, shown below. The shelves are congruent triangles. Carla measures out the first shelf, on the left, with measurements as shown. What size angle should she cut for corner *a* of the second shelf?

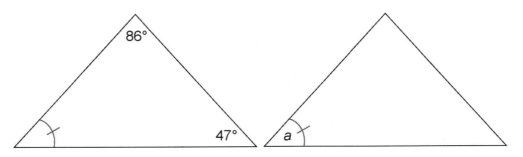

14. Ryan cuts two triangular shapes of fabric for an art project. What is the length of the side labeled with a question mark in the figure shown?

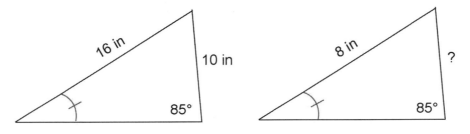

Note: Figure not drawn to scale.

15. In the figure shown, what is the measure of ∠*CDE*?

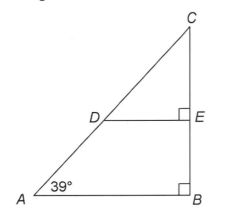

Solutions

1. The triangles are congruent based on SSS, the side-side-side test for congruence. They have three congruent sides.

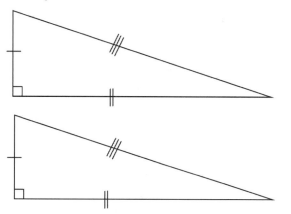

2. The triangles are similar based on AA, the angle-angle test for similarity. They have two sets of congruent corresponding angles.

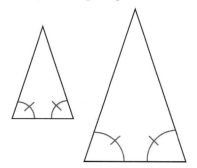

3. The triangles are not definitely congruent because they only have an SSA combination, or side-side-angle. They have two congruent sides plus one non-included angle, which does not prove congruence.

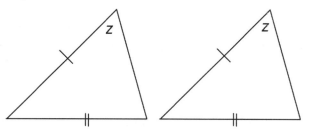

Plane Geometry

4. The triangles are similar based on SAS, the side-angle-side test for similarity. They have two sets of equally proportional corresponding sides, and the angles between these sides are congruent.

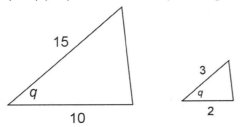

The corresponding angles labeled q are equal. The sides are in a ratio of $\dfrac{15}{3}$ and $\dfrac{10}{2}$, both of which can be reduced to 5.

5. The measurement of c is 40°. The two triangles are similar, so we know that all three sets of corresponding angles are equal.

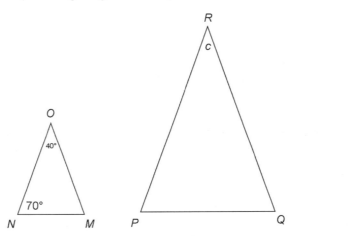

6. The measure of ∠R is 30°.

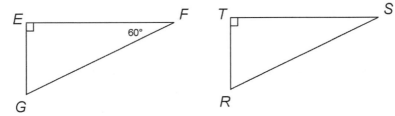

The two triangles are congruent. This means that all three corresponding sides and all three corresponding angles are equal. In this case, we know that the measure of ∠R is equal to the measure of ∠G. Use subtraction to find m∠G:

$$180° - 90° - 60° = 30°$$

The measure of ∠G is 30°. The measure of corresponding ∠R is also 30°.

7. The length of s is 6.

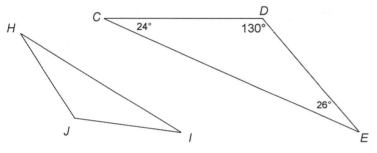

We are told that $\triangle JKL \cong \triangle ZYX$. Therefore, the triangles have three equal corresponding sides and three equal corresponding angles. We are told that $JL = 6$, so s also measures 6.

8. The measure of $\angle J$ is 130°.

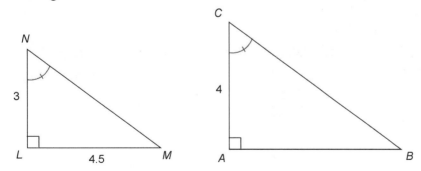

These two triangles are similar. All three angles are therefore equal. Angle J is a corresponding angle with $\angle D$, so $m \angle J = 130°$.

9. The length of $\overline{AB}$ is 6.

The triangles have two sets of equal angles. So, $\triangle NLM \sim \triangle CAB$. That means their corresponding sides are proportional, in the same ratio. To find the length of $\overline{AB}$, set up a proportion:

$$\frac{NL}{CA} = \frac{LM}{AB}$$

$$\frac{3}{4} = \frac{4.5}{AB}$$

$$3 \times AB = 4 \times 4.5$$

Cross-multiply to solve: $\quad 3AB = 18$

$$AB = 6$$

10. The measurement of $\angle E$ is 103°.

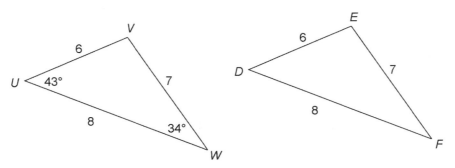

Based on the side-side-side test (SSS), the triangles are congruent. They have three sets of equal sides. That means the measure of $\angle E$ is equal to the measure of $\angle V$.

We can find the measure of $\angle V$ by subtracting the given measures from 180°:

$$180° - 43° - 34° = 103°$$

The measures of $\angle V$ and $\angle E$ both equal 103°.

11. The length of $\overline{XY}$ is 30.

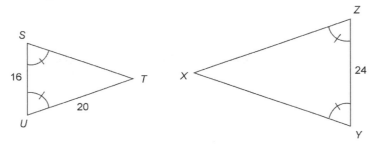

The triangles have two sets of equal angles. So, $\triangle STU \sim \triangle ZXY$. Their corresponding sides are therefore equally proportional. Set up a proportion to determine the missing length:

$$\frac{SU}{ZY} = \frac{TU}{XY}$$

$$\frac{16}{24} = \frac{20}{XY}$$

Cross-multiply to solve: $16 \times XY = 24 \times 20$

$$16XY = 480$$

$$XY = 30$$

Chapter 5: Similar and Congruent Triangles

12. The distance between buildings B and C is 100 yards.

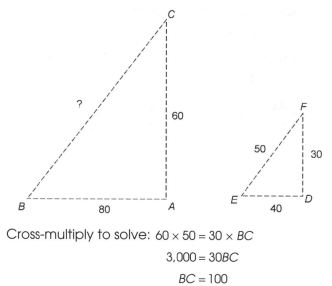

The question states that $\triangle ABC \sim \triangle DEF$. This tells us that the corresponding sides of the triangles are equally proportional.

To find the length of $\overline{BC}$, set up a proportion:

$$\frac{CA}{FD} = \frac{BC}{EF}$$

$$\frac{60}{30} = \frac{BC}{50}$$

Cross-multiply to solve: $60 \times 50 = 30 \times BC$

$$3{,}000 = 30BC$$

$$BC = 100$$

13. The measurement of $\angle a$ should be 47°.

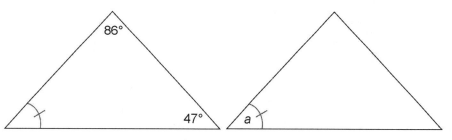

The first shelf, on the left, shows two angle measurements. To find the measure of the third angle, subtract the given angles from 180°:

$$180° - 86° - 47° = 47°$$

The measure of the third angle on the first shelf is 47°. We know that this angle is congruent to $\angle a$, because of the markings on the figure. Therefore, $\angle a$ should be cut to measure 47°.

14. The side shown with a question mark measures 5 inches.

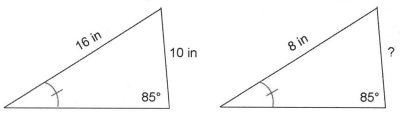

Note: Figure not drawn to scale.

Plane Geometry

The question tells us that the figure is not drawn to scale. This means we must rely only on the measurements given, and not on what the picture looks like, to determine the answer.

First, examine the triangles to see if they are congruent. They look congruent, but the measurements are not equal. The left triangle is larger than the right triangle, because 16 inches is greater than 8 inches.

Next, determine if the triangles are similar. The triangles pass the angle-angle similarity test, or AA. They have two congruent angles, as shown in the figure. So, the triangles are similar, and their corresponding sides are equally proportional.

Set up a proportion to determine the length of the missing side: $\dfrac{16}{8} = \dfrac{10}{?}$

Cross-multiply to solve:

$$16 \times \; ? = 8 \times 10$$
$$16 \times \; ? = 80$$
$$? = \frac{80}{16}$$
$$? = 5$$

The side shown with a question mark measures 5 inches.

15. The measure of $\angle CDE$ is 39°.

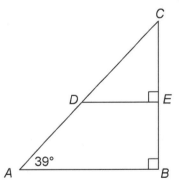

In the figure, $\triangle CAB$ is similar to $\triangle CDE$. We know this because the triangles have two equal angles: $\angle C$, which is shared, and angles CED and CBA, which are both 90°. Since the triangles are similar, all of the angles are congruent. Therefore, m$\angle CDE$ = m$\angle CAB$.

Very nice work on congruent and similar triangles. Next we look at certain types of right triangles with special properties.

Chapter 6

Special Right Triangles

Hi! I'm egghead. I will teach the following concepts in this chapter:

What is a special right triangle?	Finding angles
Common square roots	Finding lengths
45-45-90° triangles	Pythagorean triples
30-60-90° triangles	Finding lengths

What is a special right triangle?

Certain types of right triangles appear often in geometry. They are called **special right triangles.** We will look at three main types.

Three main types of special right triangles

One type of special right triangle is called the **45-45-90° triangle.** This is a right triangle with interior angles of 45, 45, and 90 degrees.

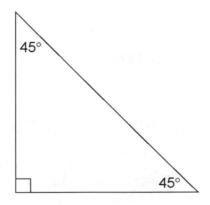

A second type of special right triangle is called the **30-60-90° triangle.** These triangles have interior angles of 30, 60, and 90 degrees.

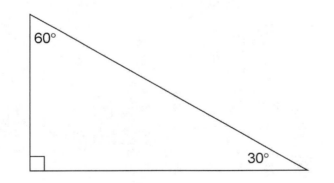

A third type of special right triangle is called the **Pythagorean triple.**

egghead's Guide to Geometry

Chapter 6: Special Right Triangles

These triangles have three sides with certain lengths, based on the Pythagorean theorem. There are four common Pythagorean triples:

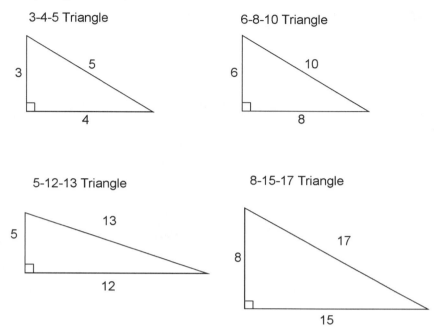

3-4-5 Triangle

6-8-10 Triangle

5-12-13 Triangle

8-15-17 Triangle

Now let's take a look at the properties of each special right triangle.

Common square roots

Before we start, there are two common square roots that you should know. These are the values for $\sqrt{2}$ and $\sqrt{3}$.

On some questions, you can leave the square root sign right in the answer. On other questions, you have to give an approximate number.

The value for $\sqrt{2}$ is approximately 1.414. It's actually a decimal number that goes on forever, but 1.414 is all you need to remember.

$$\sqrt{2} \approx 1.414$$

The value for $\sqrt{3}$ is approximately 1.732. This value is also an infinite decimal number, so 1.732 is close enough.

$$\sqrt{3} \approx 1.732$$

Use the ≈ sign when giving an estimate.

These two square roots come up often when dealing with special right triangles, so it's helpful to memorize their values.

45-45-90 triangles

The 45-45-90 triangle has two equal angles. It is an isosceles right triangle.

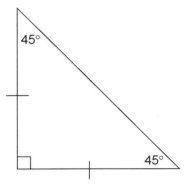

The sides across from the 45° angles are also equal. The sides of a 45-45-90 triangle always have the same ratio: $1:1:\sqrt{2}$.

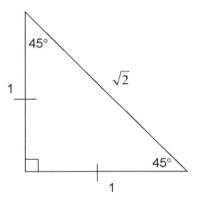

This does not mean that the sides of a 45-45-90 triangle will always measure 1, 1, and $\sqrt{2}$. It means that the sides will always be in a **ratio** of $1:1:\sqrt{2}$.

Here are some examples to show you how this works.

Examples

If one leg of a 45-45-90 triangle measures 1 inch, the other leg will measure 1 inch.

The hypotenuse will measure $\sqrt{2}$ inches.

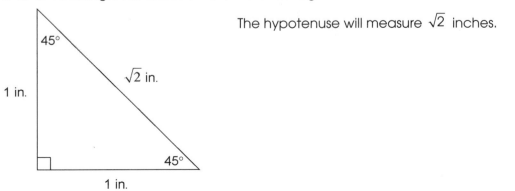

If one leg of a 45-45-90 triangle measures 2 feet, the other leg will measure 2 feet.

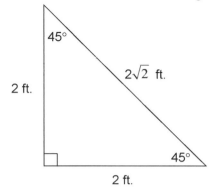

The hypotenuse will measure $2\sqrt{2}$ feet.

The sides **across** from the 45° angles always measure the same length as each other. The hypotenuse always measures the same length as one leg × $\sqrt{2}$.

30-60-90 triangles

Unlike 45-45-90 triangles, the 30-60-90 triangle has no equal angles, and it's not isosceles. However, it does have some important properties to know about.

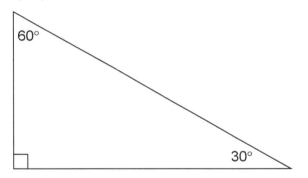

The sides of a 30-60-90 triangle are always in the ratio of $1 : \sqrt{3} : 2$.

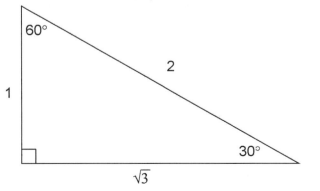

The side across from the 30° angle always has the smallest measure. The side across from the 90° angle always has the largest measure. The side across from the 60° angle has the measure in between.

Plane Geometry

This is easier to remember if we use estimation. The value for $\sqrt{3}$ is approximately 1.732, which is greater than 1 but less than 2.

Examples

If the shortest leg of a 30-60-90 triangle measures 1 foot, the other leg will measure $\sqrt{3}$ feet.

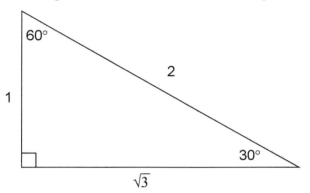

The hypotenuse will measure 2 feet.

If the shortest leg of a 30-60-90 triangle measures 4 inches, the other leg will measure $4\sqrt{3}$ inches.

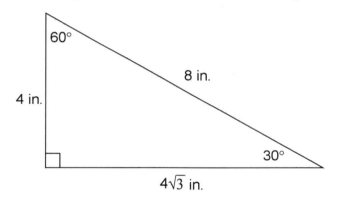

The hypotenuse will measure 8 inches.

If the shortest leg of a 30-60-90 triangle measures 7 meters, the other leg will measure $7\sqrt{3}$ meters.

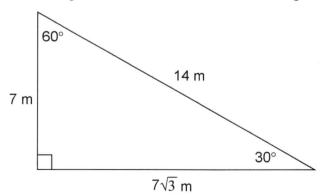

The hypotenuse will measure 14 meters.

The side **across** from the 60° angle always measures $\sqrt{3}$ times the length of the shortest leg. The hypotenuse always measures 2 times the shortest leg.

Finding angles

We can use the properties of 45-45-90 and 30-60-90 triangles to find the measure of angles and lengths. What is the measure of y in the figure shown?

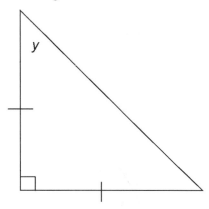

This is a 45-45-90 triangle. We know this because it has two equal sides. The two sides lie opposite the equal angles, so y must measure 45°.

What is the measure of z in the figure shown?

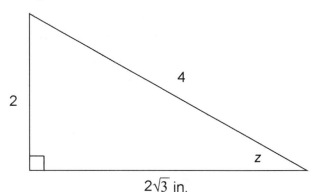

$2\sqrt{3}$ in.

This is a 30-60-90 triangle. We know this because the sides are in a ratio of $2 : 2\sqrt{3} : 4$. If we divide each of the side lengths by 2, this gives us the ratio of $1 : \sqrt{3} : 2$.

The side marked 2 is the shortest side of the triangle, and $\angle z$ lies opposite the shortest side. So, z must measure 30°.

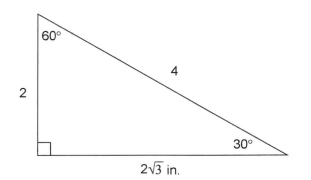

What is the measure of *s* in the figure shown?

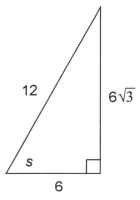

This is a 30-60-90 triangle, with sides in a ratio of $6 : 6\sqrt{3} : 12$. If we divide each of the side lengths by 6, this reduces to a ratio of $1 : \sqrt{3} : 2$.

The side marked 6 is the shortest side of the triangle, and the side marked $6\sqrt{3}$ is the next largest side. Angle *s* lies opposite of the side marked $6\sqrt{3}$, so *s* must measure 60°.

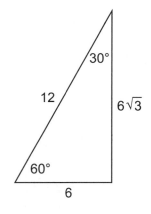

Finding lengths

Triangle properties can help us find side lengths also. What is the length of the other leg in the figure shown?

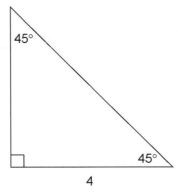

This is a 45-45-90 triangle, so both legs have the same measurement. The missing leg also measures 4.

The length of the hypotenuse can be determined as well. Its measure is $4\sqrt{2}$.

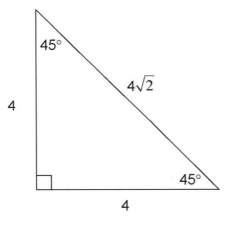

What are the measures of j and k in the figure shown?

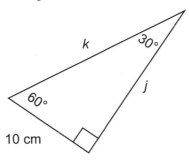

Plane Geometry

This is a 30-60-90 triangle, so the sides are in the ratio of $1 : \sqrt{3} : 2$. We are told that the shortest leg measures 10 centimeters. So, side *j* must measure $10\sqrt{3}$ centimeters, and *k* must measure 20 centimeters.

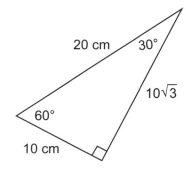

What are the lengths of $\overline{AB}$ and $\overline{AC}$ in the figure shown?

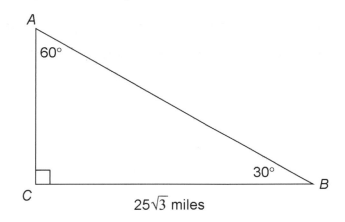

We are told that $CB = 25\sqrt{3}$. This is the side across from the 60° angle. Side $\overline{AC}$, across from the 30° angle, must measure 25.

The length of the hypotenuse $\overline{AB}$ can be determined as well. Its measure is 2 times the shortest side, or 50.

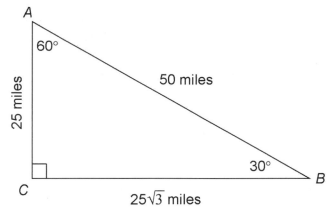

Practice Questions

1. What is the measure of ∠*DFE* in the figure shown?

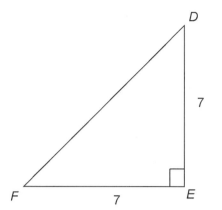

2. In the figure shown, what is the measure of $\overline{DF}$?

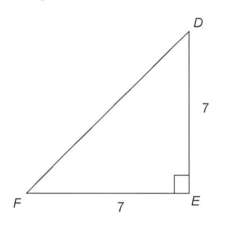

3. In the figure shown, $NP = 5\sqrt{2}$ centimeters. What is the length, in centimeters, of $\overline{OP}$?

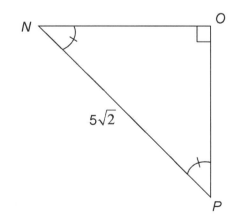

4. What is the measure of ∠*QRS* in the figure shown?

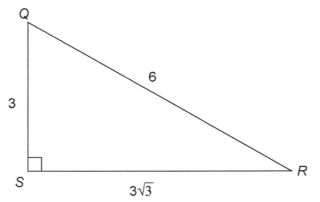

5. In the figure shown, what is the measure of $\overline{XY}$?

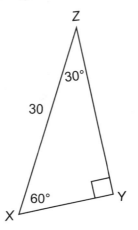

6. In the figure shown, what is the measure of *s*?

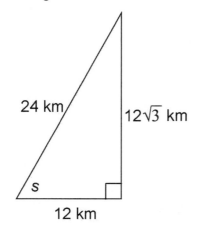

Solutions

1. The triangle has two equal sides and one right angle, so it is a 45-45-90 triangle. The two smaller angles are equal. Therefore, the measure of ∠*DFE* is 45°.

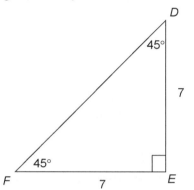

2. Triangle *DEF* is a 45-45-90 triangle, and both legs measure 7. The hypotenuse is equal to leg × $\sqrt{2}$. Therefore, the measure of $\overline{DF}$ is $7\sqrt{2}$.

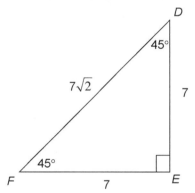

3. This right triangle has two equal angles, so it is a 45-45-90 triangle. The hypotenuse is equal to leg × $\sqrt{2}$. We are told that *NP* = $5\sqrt{2}$ centimeters, so both legs measure 5 centimeters.

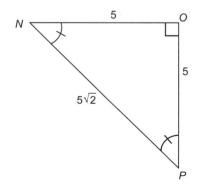

4. The measure of ∠QRS is 30°.

The sides of this triangle are in a ratio of $3 : 3\sqrt{3} : 6$. If we divide each of the side lengths by 3, this reduces to a ratio of $1 : \sqrt{3} : 2$. So, we have a 30-60-90 triangle.

The side marked 3 is the shortest side of the triangle, and ∠QRS lies opposite this shortest side. That tells us that m∠QRS = 30.

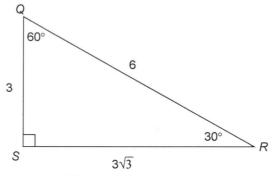

5. The length of $\overline{XY}$ is 15.

This is a 30-60-90 triangle, so the sides are in the ratio of $1 : \sqrt{3} : 2$. The hypotenuse measures 30, and the shortest side measures half of that. Therefore, $\overline{XY}$ measures 15.

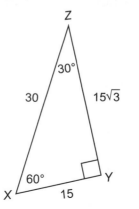

6. The measure of *s* is 60°.

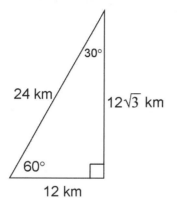

The sides of this triangle are in a ratio of $12 : 12\sqrt{3} : 24$. We can reduce this to $1 : \sqrt{3} : 2$ if we divide each side by 12. So, this is a 30-60-90 triangle, with 12 kilometers as the shortest side.

The angle marked *s* is across from the second shortest side, so it must measure 60°.

Pythagorean triples

The third type of special right triangle that we'll review is called the **Pythagorean triple.** Pythagorean triples are right triangles with side lengths that are positive integers.

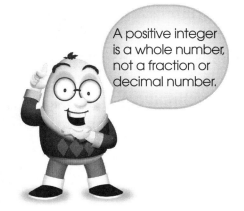

A positive integer is a whole number, not a fraction or decimal number.

Examples

There are an infinite number of Pythagorean triples. In this chapter, we'll look at four of the most common.

3-4-5 triangles

The smallest Pythagorean triple is the 3-4-5 triangle. This is a right triangle whose side lengths measure 3, 4, and 5:

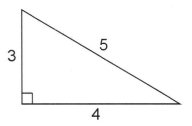

If you use the Pythagorean theorem to check these side lengths, you'll see they fit perfectly:

$$a^2 + b^2 = c^2$$
$$3^2 + 4^2 = 5^2$$
$$9 + 16 = 25$$
$$25 = 25$$

6-8-10 triangles

Another Pythagorean triple is the 6-8-10 triangle. The legs of this triple are multiples of the 3-4-5 triangle. Take each side of the 3-4-5 triangle and multiply it by 2. This gives you a triangle with side lengths 6, 8, and 10.

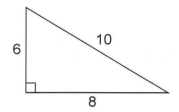

Other triplets in this series include 9-12-15 and 12-16-20.

5-12-13 triangles

A third common Pythagorean triple is the 5-12-13 triangle. This triangle has legs of length 5, 12, and 13. Other multiples are 10-24-26 and 15-36-39.

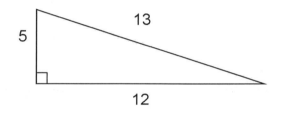

8-15-17 triangles

The fourth common Pythagorean triple is the 8-15-17 triangle. Again, there are many multiples of this, but 8-15-17 is most important to remember.

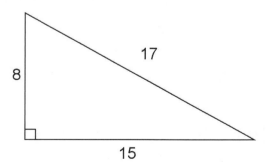

Finding lengths

Knowing some Pythagorean triples gives you a shortcut to finding the lengths of the sides of certain right triangles. When you recognize a triple, you don't have to use the Pythagorean theorem to find the length of a missing side.

For example, in the figure below, we don't have to calculate to find the value of c. This is a right triangle with legs of length 3 and 4. It is a 3-4-5 Pythagorean triple, so c must measure 5.

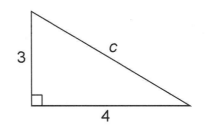

Similarly, in the figure below, we can find the length of a just by recognizing the Pythagorean triple. This is a 6-8-10 Pythagorean triple. So, the length of a is 6.

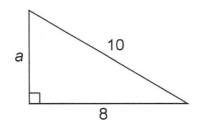

Practice Questions

1. What is the length of b in the figure shown?

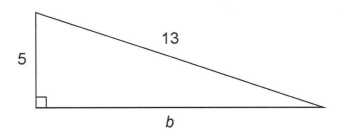

2. What is the length of $\overline{RT}$ in the figure shown?

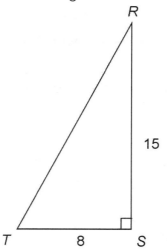

3. Find the value of x in the figure shown.

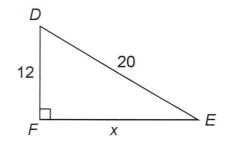

4. Find the length of the missing side in the figure shown.

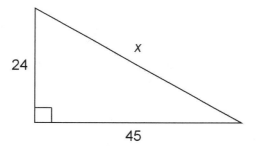

5. Find the length of the missing side.

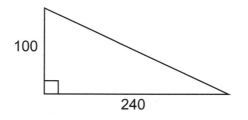

egghead's Guide to Geometry

6. Find the length of the missing side.

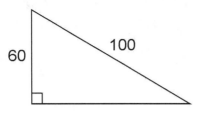

 **Solutions**

1. The length of *b* is 12. This Pythagorean triple has sides of length 5, 12, and 13. So, the missing side (side *b*) must measure 12.

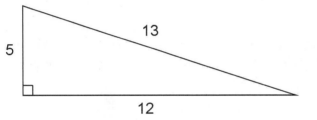

2. The length of $\overline{RT}$ is 17. This is an 8-15-17 Pythagorean triple. The hypotenuse measures 17.

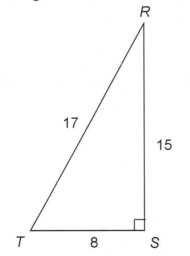

3. The value of *x* is 16. You are given the lengths of two sides, 12 and 20. Both of these sides are multiples of a 3-4-5- triangle. The lengths are 4 times as long as the 3-4-5 lengths. For leg $\overline{DF}$, $3 \times 4 = 12$. For hypotenuse $\overline{DE}$, $5 \times 4 = 20$. Side $\overline{FE}$ must also be a multiple: $4 \times 4 = 16$.

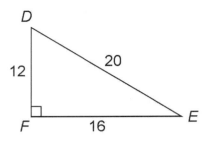

4. This is a multiple of an 8-15-17 triangle. The lengths of the sides are three times as long as the sides of an 8-15-17 triangle. If we multiply 8 by 3, we get 24. If we multiply 15 by 3, we get 45. The missing length is therefore 3×17, or 51.

5. This is a multiple of a 5-12-13 triangle. The lengths 100 and 240 are 20 times the lengths of the corresponding sides of a 5-12-13 triangle. So the missing length is 20×13, or 260.

6. This triangle is a multiple of a 6-8-10 Pythagorean triple. The sides marked 60 and 100 are 10 times the lengths of the corresponding sides of a 6-8-10 triangle. The missing length is therefore 8×10, or 80.

Chapter Review

1. What is the measure of ∠*LMN* in the figure shown?

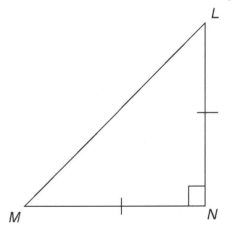

2. What is the measure of hypotenuse $\overline{RS}$ in the figure shown?

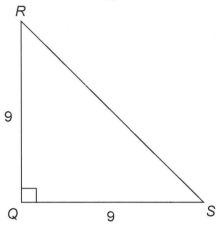

3. What is the length of $\overline{FH}$ in the figure shown?

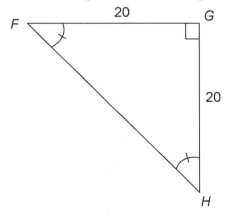

4. Jordan's house is the same distance from the gym as it is from his school. He lives 3 miles from both the gym and school, as shown. What is the distance, in miles, from the gym to the school?

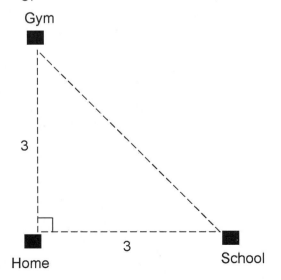

5. Aimee folds a napkin in the dimensions shown. The top corner of the napkin forms a right triangle. What is the length, in inches, of *n*?

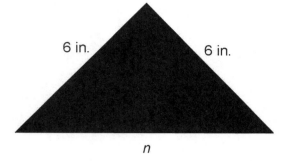

6 in.   6 in.

n

6. What is the measure of ∠*KLM* in the figure shown?

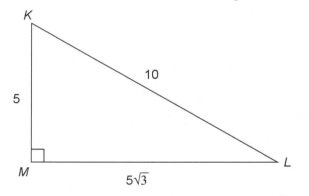

K

10

5

M

L

$5\sqrt{3}$

7. In the figure shown, what is the measure of $\overline{PQ}$?

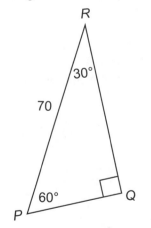

R

30°

70

60°   *Q*

P

8. What is the measure of *w* in the figure shown?

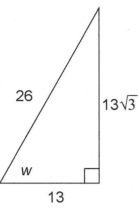

9. Allen delivers packages to delivery stops *B*, *C*, and *D*, shown in the figure below. He starts at stop *C* and drives 19 miles north to stop *D*. He then drives 38 miles southeast to stop *B*. He realizes he forgot to get a delivery signature at stop *C*, so he drives back to stop *C* from stop *B*. How many miles must he travel due west to return to stop *C*?

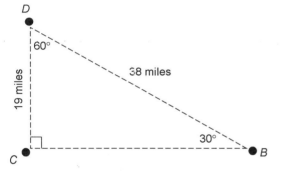

10. In the figure shown, what is the measure of $\overline{VW}$?

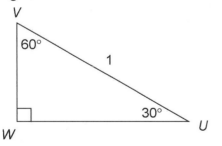

11. What is the measure of $\overline{UW}$ in the figure shown?

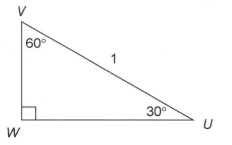

12. What is the length of *m* in the figure shown?

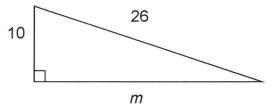

13. What is the length of $\overline{CD}$ in the figure shown?

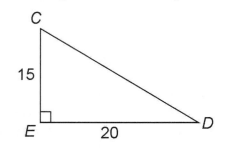

14. Find the value of *h* in the figure shown.

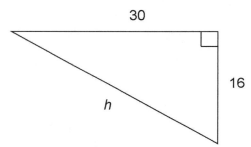

15. Brandy cuts two pieces of construction paper for a collage. The pieces of paper are shaped like triangles, as shown. What are the lengths of *x* and *y*?

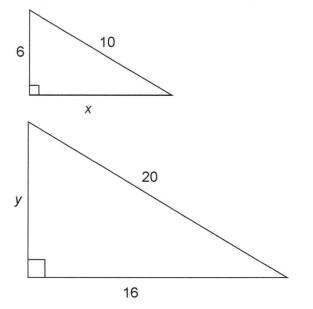

Chapter 6: Special Right Triangles

Solutions

1. Triangle *LMN* has two equal sides and one right angle, so it is a 45-45-90 triangle. The two smaller angles are equal. Therefore, the measure of ∠*LMN* is 45°.

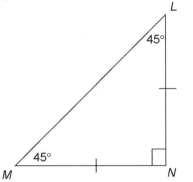

2. Triangle *QRS* is a 45-45-90 triangle, with two legs of length 9. The hypotenuse $\overline{RS}$ is equal to leg × $\sqrt{2}$. Therefore, $\overline{RS}$ measures $9\sqrt{2}$.

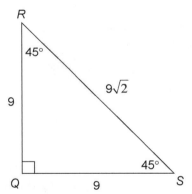

3. Triangle *FGH* is a 45-45-90 triangle. We know this because it has two equal angles and two equal sides. The sides of a 45-45-90 triangle are always in a ratio of $1:1:\sqrt{2}$. Hypotenuse $\overline{FH}$ is therefore equal to $20\sqrt{2}$.

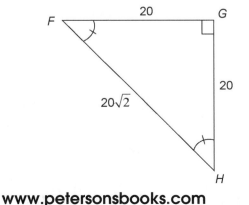

4. The distance from the gym to the school is $3\sqrt{2}$ miles. The three buildings form an isosceles right triangle, as shown in the figure. The two legs of the triangle measure 3. The smaller angles both measure 45°.

The sides of a 45-45-90 triangle are always in a ratio of $1:1:\sqrt{2}$. That means the hypotenuse equals leg $\times \sqrt{2}$. So the distance from the gym to the school, the hypotenuse of the triangle, is $3\sqrt{2}$.

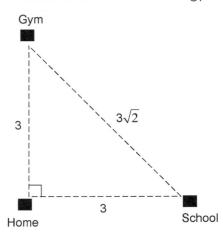

5. The length of n is $6\sqrt{2}$ inches.

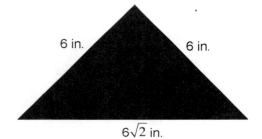

The napkin forms an isosceles right triangle. The side across from the 90° angle, the hypotenuse, is equal to the leg $\times \sqrt{2}$. The hypotenuse therefore measures $6\sqrt{2}$ inches, or approximately 8.48 inches.

6. The measure of $\angle KLM$ is 30°.

The sides of this triangle are in a ratio of $5:5\sqrt{3}:10$. If we divide each of the side lengths by 5, this reduces to a ratio of $1:\sqrt{3}:2$. This means KLM is a 30-60-90 triangle.

The side marked 5 is the shortest side of the triangle, and $\angle KLM$ lies opposite this shortest side. So, we know that m $\angle KLM = 30$.

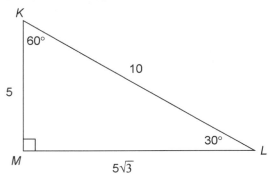

7. The length of $\overline{PQ}$ is 35.

This is a 30-60-90 triangle with sides in the ratio of $1 : \sqrt{3} : 2$. The hypotenuse measures 70, and the shortest side measures half of 70. Therefore, $\overline{PQ}$ measures 35.

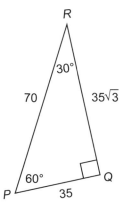

8. The measure of *w* is 60°.

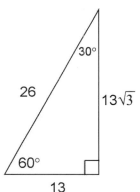

This triangle has sides in a ratio of $13 : 13\sqrt{3} : 26$. We can reduce this to $1 : \sqrt{3} : 2$ if we divide each side by 13. This is therefore a 30-60-90 triangle.

We are asked for the measure of *w*, which lies across from the second shortest side. The measure of *w* must be 60°.

9. Allen must travel $19\sqrt{3}$ miles to return to stop *C*.

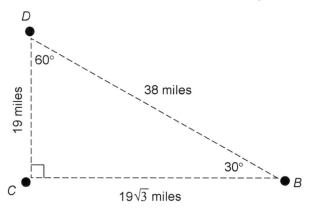

The triangle created by Allen's path is a 30-60-90 triangle. We are given the lengths of two of the sides: 19 miles (for the shortest leg) and 38 miles (for the hypotenuse). The side across from the 60° angle is equal to the length of the shortest leg $\times \sqrt{3}$, which in this case equals $19\sqrt{3}$ miles.

10. The measure of $\overline{VW}$ is $\frac{1}{2}$. It can also be written as a decimal number.

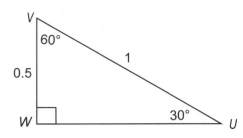

Every 30-60-90 triangle has sides in a ratio of $1 : \sqrt{3} : 2$. Triangle *UVW* is no exception. In this case, the longest side measures 1. So, the shortest side must measure half of that, or 0.5.

11. The measure of $\overline{UW}$ is $\frac{1}{2}\sqrt{3}$. It can also be written as a decimal number.

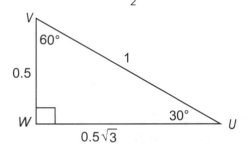

In this triangle, side $\overline{UW}$ lies opposite the angle measuring 60°. So, the measure of side $\overline{UW}$ will equal the length of the shortest leg times $\sqrt{3}$. The shortest leg measures $\frac{1}{2}$, so $\overline{UW}$ measures $\frac{1}{2}\sqrt{3}$ in fractional form. In decimal form, this is written $0.5\sqrt{3}$.

12. The length of *m* is 24. The triangle is a multiple of the 5-12-13 Pythagorean triple. Each side is twice as long as the side of a 5-12-13 triangle. So, the missing side *m* measures 2 × 12, or 24.

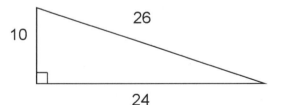

13. The length of $\overline{CD}$ is 25. This triangle is a multiple of the 3-4-5 Pythagorean triple. Each side is 5 times as long as the 3-4-5 triangle sides. Hypotenuse $\overline{CD}$ therefore measures 5 × 5, or 25.

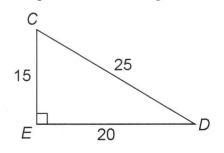

14. The value of *h* is 34.

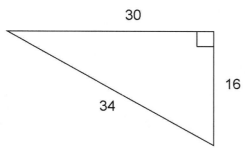

This triangle is a multiple of the 8-15-17 Pythagorean triple. Each side is twice as long as the 8-15-17 triangle sides. The length of *h*, the hypotenuse, is therefore 2 × 17, or 34.

15. The length of *x* is 8, and the length of *y* is 12.

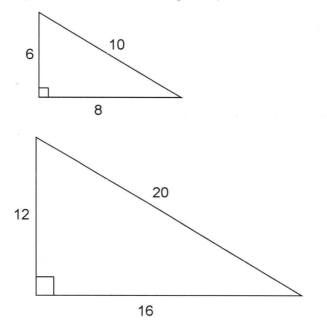

The triangles are both Pythagorean triples. The first is a 6-8-10 triangle, so side *x* measures 8. The larger triangle has sides exactly twice as long as the sides of the 6-8-10 triangle. Multiply each side of the smaller triangle by 2 to get the side lengths of the larger triangle.

The missing leg, *y*, measures 2 × 6, or 12.

Chapter 7

Circles

Hi! I'm egghead. I will teach the following concepts in this chapter:

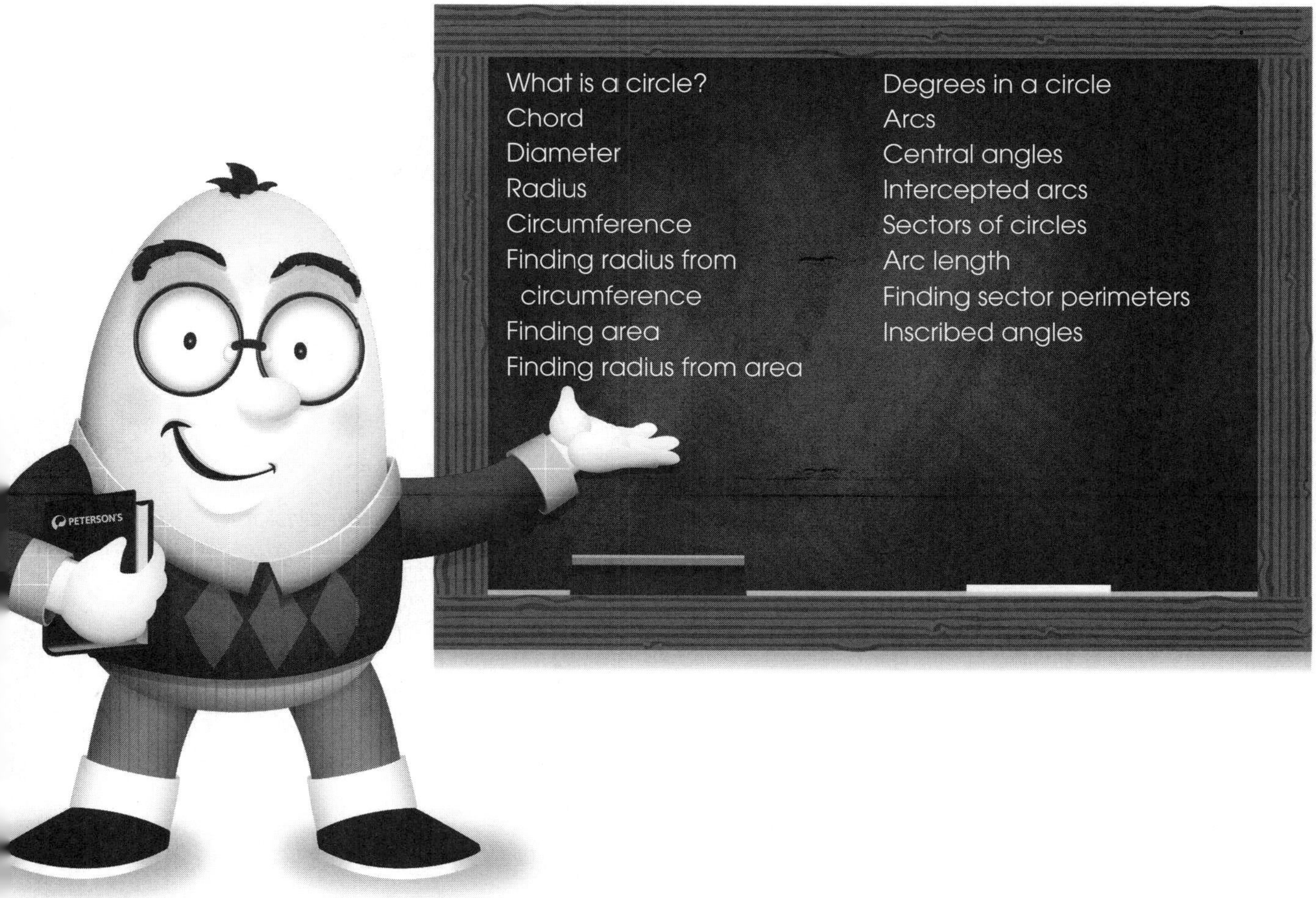

What is a circle?	Degrees in a circle
Chord	Arcs
Diameter	Central angles
Radius	Intercepted arcs
Circumference	Sectors of circles
Finding radius from circumference	Arc length
Finding area	Finding sector perimeters
Finding radius from area	Inscribed angles

What is a circle?

We all know what a circle looks like. It's a perfectly round shape that looks like this:

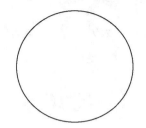

Every circle has a center. Point *O* lies exactly in the center of the circle shown:

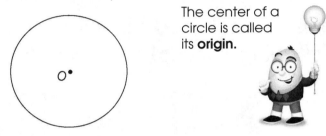

The center of a circle is called its **origin.**

To name a circle, use its center point. The figure above shows circle *O*.

Every point on the circle is the same distance from the center of the circle:

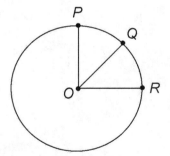

Just like the other shapes we've studied, a circle also lies in a two-dimensional plane.

In the diagram below, circle *O* lies in a plane. Points *P*, *Q*, and *R* lie on the circle.

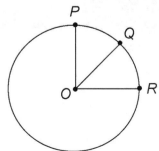

Chord

A **chord** is a line segment that connects any two points on a circle. In the figure below, you can see chord *AB*.

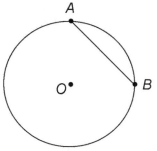

Diameter

One important measurement of a circle is a special chord called the **diameter.** The diameter is the distance from one side of a circle to the other side:

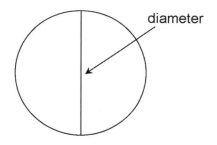

The diameter measures the distance *across* the circle. It goes through the center of the circle:

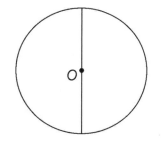

This is the diameter of circle *O*.

You could also draw the diameter across the circle, like this:

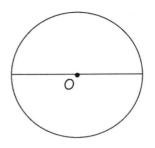

It really doesn't matter where you draw the diameter. It just has to go through the center of the circle.

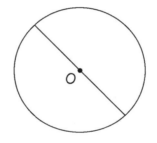

Examples

The diameter of this circle measures 5.

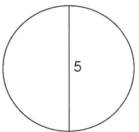

The diameter of this circle measures 9.

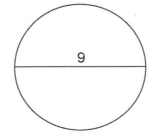

egghead's Guide to Geometry

This is circle *P* with diameter *AB*.

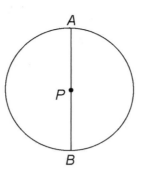

The bigger the circle, the bigger the diameter.

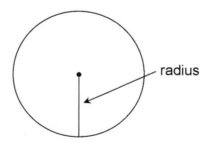

Diameter *YZ* has length 20.

Radius

The radius of a circle is the distance from the center of the circle to its edge:

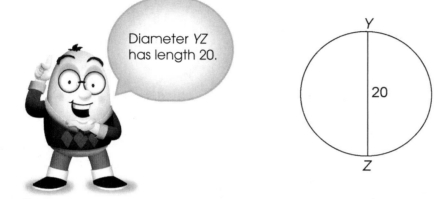

radius

The radius starts at the center of the circle and extends to the outside edge:

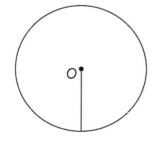

This is a radius of circle *O*

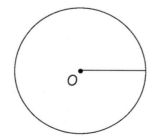

This is also a radius of circle *O*.

Plane Geometry

It doesn't matter where you draw the radius of a circle. It just has to start at the center of the circle and extend to the outside edge.

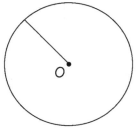

Examples

The radius of this circle measures 2:

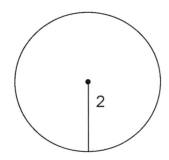

This is circle *P* with radius *PA* and radius *PB*:

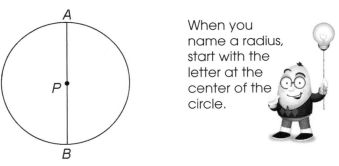

When you name a radius, start with the letter at the center of the circle.

In circle *X*, radius *XY* has length 10. Radius *XZ* has length 10, too.

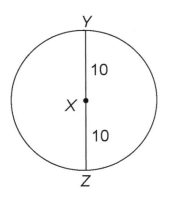

The radius is exactly $\frac{1}{2}$ the diameter. In circle X, diameter YZ has length 20.

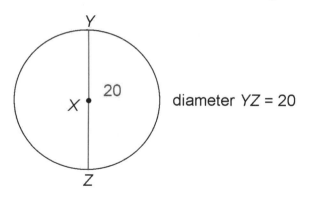

diameter $YZ = 20$

Practice Questions

1. In the space below, draw a circle and label it Q.

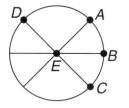

2. What is the name of this circle?

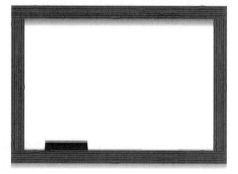

3. In circle O shown, what is the length of radius OK?

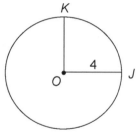

4. In circle *C* shown, what is the length of radius *CD*?

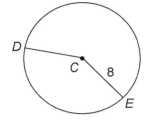

5. In the circle below, what is the length of the diameter?

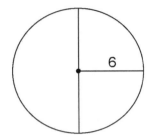

Solutions

1. The correct answer is shown below.

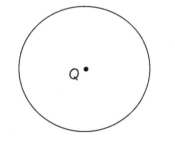

2. Though there are many chords in this circle, the center point of the circle is *E*. So the circle is named circle *E*.

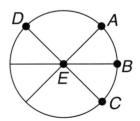

3. In circle *O* shown, the length of radius *OK* is also 4. All lines that extend from the center of the circle to its edge have the same measurements.

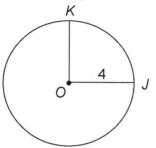

4. The figure tells us that radius *CE* measures 8. Every radius on a circle has the same length. So, radius *CD* must also measure 8.

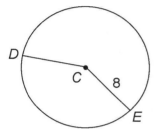

5. In the circle below, we know the radius is 6. The diameter is double the radius, or 12.

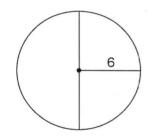

Circumference

As we learned in earlier chapters, perimeter is the distance around a shape.

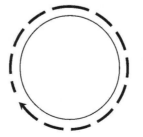

The perimeter of a circle is the distance around its outside edge. This is also called the **circumference.**

Plane Geometry

The circumference of this circle is 45:

circumference = 45

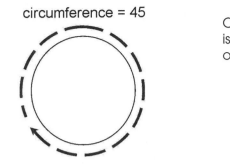

Circumference is the perimeter of a circle.

The circumference of this circle is 22:

circumference = 22

The circumference of this larger circle is 130:

circumference = 130

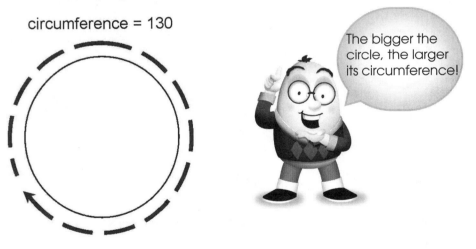

The bigger the circle, the larger its circumference!

Formula

To find the circumference of a circle, we use this formula:

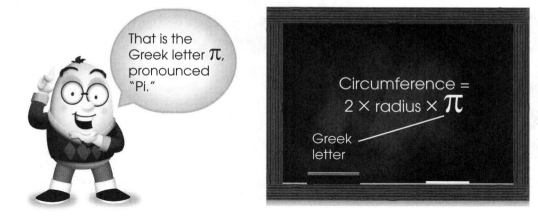

That is the Greek letter π, pronounced "Pi."

Circumference = 2 × radius × π

Greek letter

The Greek letter π is a symbol for a certain number. The number is equal to about 3.14. We say "about" because π is actually a decimal number that extends into infinity. The long version looks like this:

$$\pi = 3.14159265359\ldots$$

decimals keep going

For short, we say Pi equals approximately 3.14:

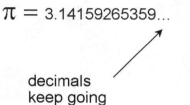

$$\pi \approx 3.14$$

The wavy equals sign means "approximately."

The story of how π was found is very complex. Many scholars contributed to it. But the Greeks definitely had a lot to do with it, because it ended up with a Greek name!

Some brilliant folks figured out that the circumference of any circle can be found by multiplying π by twice the radius.

Any circle?

Yes! Every one!

It turns out that π is a magic number in geometry. It helps you find the answers to certain questions. Math is cool like that.

The formula for circumference, again, is:

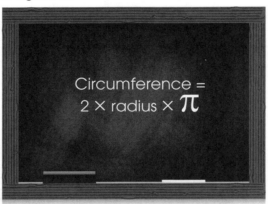

Circumference =
2 × radius × π

Here's the short form:

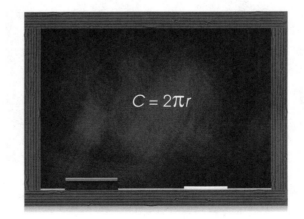

$C = 2\pi r$

Examples

Many geometry circle questions ask you to find the circumference.

Here are some examples.

This circle has a radius of length 2:

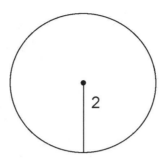

To find the circumference, multiply 2 × radius × π. The radius is 2, so we multiply 2 × 2 × π, or 2 × 2π. The circumference is 4π.

It's okay to leave the symbol π in your answer!

This circle has a radius of length 3:

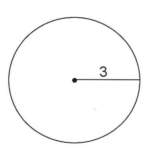

To find the circumference, multiply 2 × radius × π. This time, the radius is 3. So, we multiply 2 × 3 × π. The circumference is 6π.

Finding radius from circumference

If you know the circumference of a circle, you can also find the radius.

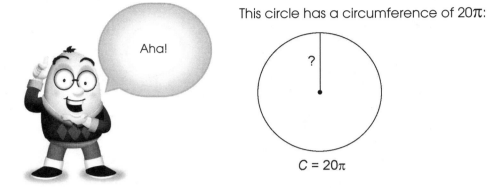

This circle has a circumference of 20π:

$C = 20\pi$

To find the radius, divide the circumference by 2π. The circumference is 20π, so we divide 20π by 2π. The radius is 10.

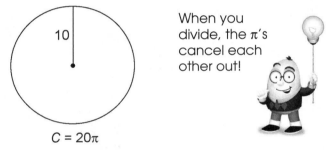

$C = 20\pi$

When you divide, the π's cancel each other out!

Here's another example. This circle has a circumference of 150π:

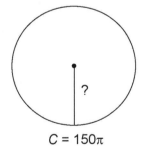

$C = 150\pi$

To find the radius, divide the circumference by 2π. The circumference is 150π, so we divide 150π by 2π. The radius is 75.

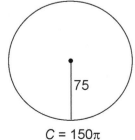

$C = 150\pi$

Practice Questions

1. What is the circumference of circle *O*?

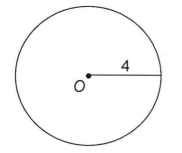

2. Find the circumference of circle *P*.

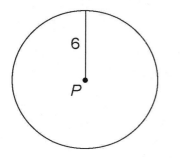

3. Cathy works at the pizza restaurant Little Davey's. She makes about 15 pizzas an hour in the restaurant's huge wood oven. Each pizza she makes has a radius of 6 inches. What's the total circumference of the 15 pizzas Cathy makes each hour?

4. Johnny and Sally love going to the ice skating rink in the winter. This winter, Sally skated halfway across the circular rink, or 7 feet from the center of the rink to its outside edge. On his first pass, Johnny skated across the entire rink. On his second pass, he skated around the entire rink. How far did Johnny skate on his first two passes combined?

5. In December 1990, a group of folks got together in Norwood, South Africa and constructed the largest pizza ever made up until that time, according to the Guinness Book of World Records. The pizza measured 37.4 meters in diameter. What were its radius and its circumference?

Solutions

1. The circumference is 8π.

To find the circumference, multiply $2 \times$ radius $\times \pi$. The radius of circle O is 4. So, we multiply $2 \times 4 \times \pi$:

$$2 \times 4 \times \pi = 8\pi$$

2. To find the circumference, multiply $2 \times$ radius $\times \pi$. The radius of circle P is 6. So, we multiply $2 \times 6 \times \pi$. The circumference is 12π.

3. To find the solution, let's first figure out the circumference of one of Cathy's pizzas. Each pizza has a radius of 6 inches. To find the circumference of one pizza, we multiply $2 \times 6 \times \pi = 12\pi$. The total circumference of the 15 pizzas must be 180π inches.

4. If Sally skated 7 feet and that measured halfway across the rink, then the rink's radius is 7 feet. So the circumference is $7 \times 2 \times \pi = 14\pi$. But Johnny also skated twice as far as Sally, so we need to add another 14 feet. Johnny skated a total of $14\pi + 14$ feet.

5. If the diameter of the pizza was 37.4 m, then its radius is $\frac{1}{2}$ that, or 18.7 m. Its circumference is just $37.4 \times \pi = 37.4\pi$ meters.

Finding area

Along with circumference, the next major topic to know about circles is how to find the area.

The area of a circle is the amount of space the circle covers.

The dotted line shows the circumference of the circle. The shaded section shows the area:

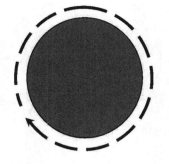

The larger the circle, the more space it takes up.

Larger circles have larger areas!

Formula

To find the area of a circle, we use this formula:

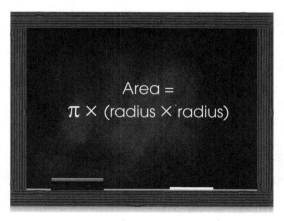

Area =
π × (radius × radius)

Yes—another cool math concept. If you multiply the radius of any circle by itself, and then multiply that by π, you get the area.

Plane Geometry

Here's the area formula in shorthand:

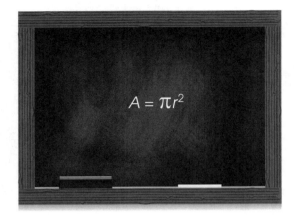

$$A = \pi r^2$$

Examples

Let's look at a couple of examples.

This circle has a radius of length 2 units.

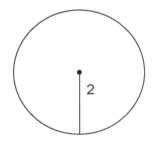

To find the area, multiply $\pi \times$ radius $\times$ radius. The radius is 2, so we multiply $\pi \times 2 \times 2$. The area is 4π square units.

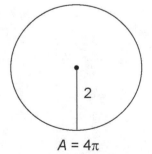

$$A = 4\pi$$

This circle has a radius of length 8 feet.

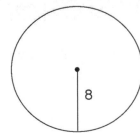

To find the area, multiply π × radius × radius. The radius is 8, so we multiply π × 8 × 8. The area is 64π square feet.

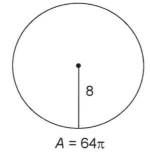

$A = 64\pi$

Practice Questions

1. What is the area of circle *X*?

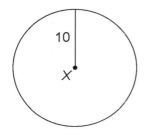

2. Find the area of circle *R*.

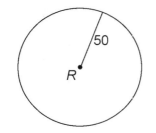

3. Remember the largest pizza ever made in 1990, according to the Guinness Book of World Records? The pizza had a diameter measuring 37.4 meters. What was its area?

4. Find the area of the circle below.

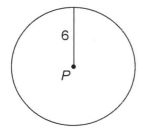

5. Find the area of the circle shown.

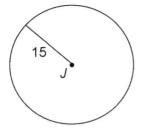

Solutions

1. To find the area, multiply π × radius × radius. The radius of circle X is 10. So, we multiply π × 10 × 10. The area is 100π square units.

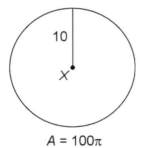

$A = 100\pi$

2. To find the area, multiply π × radius × radius. The radius of circle R is 50. So, we multiply π × 50 × 50. The area is $2,500\pi$ square units.

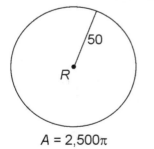

$A = 2,500\pi$

3. To find the area, let's first find the radius. Divide 37.4 by 2 and you get $r = 18.7$ meters. For the area, multiply π × r × r, or 18.7 × 18.7 × π = 349.69π square meters.

4. The area is 36π square units.

The radius of circle *P* is 6. So, we multiply π × 6 × 6.

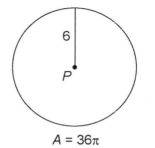

$A = 36\pi$

5. The circumference is 225π square units.

The radius of circle *J* is 15. So, we multiply π × 15 × 15.

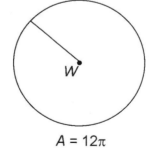

$A = 225\pi$

Finding radius from area

When you're given the area of a circle, you can also find its radius. Just work backward from the information given.

Examples

Circle *W* shown below has an area of 12π. What is the length of its radius?

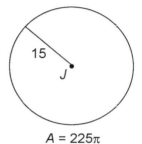

$A = 12\pi$

To answer this question, start with the area formula. Substitute 12π for the value of the area. Then solve for the radius, *r*.

$$A = \pi r^2$$

$$12\pi = \pi r^2$$

Divide both sides by π:

$$12\pi = \pi r^2$$

$$\frac{12\pi}{\pi} = \frac{\pi r^2}{\pi}$$

$$12 = r^2$$

Plane Geometry

The π's drop out, and we are left with $12 = r^2$. To find r, take the square root of both sides:

$$r^2 = 12$$
$$\sqrt{r^2} = \sqrt{12}$$
$$\sqrt{r \times r} = \sqrt{12}$$
$$r = \sqrt{12}$$

The square root of 12 can be broken down even further. The number 12 is the product of 4 times 3. So, $\sqrt{12} = \sqrt{4 \times 3}$. This is equal to $\sqrt{2 \times 2 \times 3}$. The square root of 2 times 2 is 2, so take the number 2 out from under the square root sign. The correct answer is $2\sqrt{3}$.

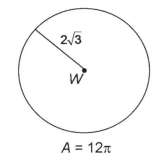

$$A = 12\pi$$

Here's another example.

Circle E below has an area of 32π square inches. To find the length of its radius, start with the area formula.

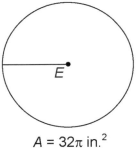

$$A = 32\pi \text{ in.}^2$$

We know that the area is 32π square inches. Plug that into the formula:

$$A = \pi r^2$$
$$32\pi = \pi r^2$$

Next, divide both sides by π. This leaves $32 = r^2$.

$$32\pi = \pi r^2$$
$$\frac{32\pi}{\pi} = \frac{\pi r^2}{\pi}$$
$$32 = r^2$$

Chapter 7: Circles

To solve for r, take the square root of both sides:

$$r^2 = 32$$
$$\sqrt{r^2} = \sqrt{32}$$
$$\sqrt{r \times r} = \sqrt{32}$$
$$r = \sqrt{32}$$
$$= \sqrt{16 \times 2}$$
$$= \sqrt{4 \times 4 \times 2}$$
$$= 4\sqrt{2}$$

The radius is $4\sqrt{2}$ inches.

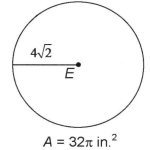

$A = 32\pi$ in.^{2}

Practice Questions

1. What is the radius of circle M?

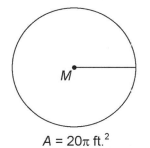

$A = 20\pi$ ft.^{2}

2. Find the diameter of circle J.

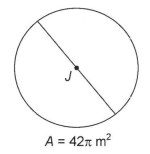

$A = 42\pi$ m^2

Plane Geometry

3. Circle *Y* has an area of 63π square yards. What is length of $\overline{BC}$?

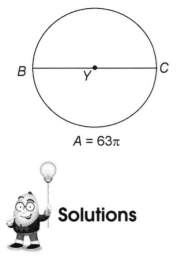

$A = 63\pi$

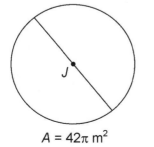 **Solutions**

1. The radius measures $2\sqrt{5}$ feet. The area of circle *M* is 20π square feet. Plug this into the area formula:

$$A = \pi r^2$$
$$20\pi = \pi r^2$$
$$\frac{20\pi}{\pi} = \frac{\pi r^2}{\pi}$$
$$20 = r^2$$

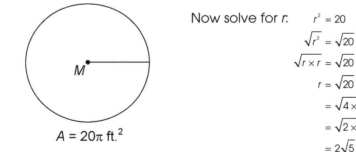

Now solve for *r*.

$$r^2 = 20$$
$$\sqrt{r^2} = \sqrt{20}$$
$$\sqrt{r \times r} = \sqrt{20}$$
$$r = \sqrt{20}$$
$$= \sqrt{4 \times 5}$$
$$= \sqrt{2 \times 2 \times 5}$$
$$= 2\sqrt{5}$$

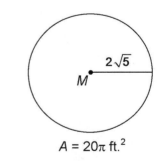

$2\sqrt{5}$

$A = 20\pi$ ft.^{2}

$A = 20\pi$ ft.^{2}

2. The diameter measures $2\sqrt{42}$ meters.

Start with the area formula:

$$A = \pi r^2$$
$$42\pi = \pi r^2$$
$$\frac{42\pi}{\pi} = \frac{\pi r^2}{\pi}$$
$$42 = r^2$$

$A = 42\pi$ m^2

egghead's Guide to Geometry

Next, solve for r.

$$r^2 = 42$$
$$\sqrt{r^2} = \sqrt{42}$$
$$\sqrt{r \times r} = \sqrt{42}$$
$$r = \sqrt{6 \times 7}$$
$$= \sqrt{42}$$

The number 42 cannot be broken down into any perfect squares, so we just leave 42 under the square root sign. The length of the diameter is twice the length of the radius. So, $d = 2\sqrt{42}$ meters.

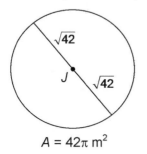

$A = 42\pi \text{ m}^2$

3. The length of $\overline{BC}$ is $6\sqrt{7}$ yards.

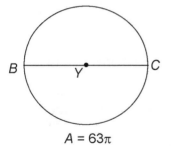

$A = 63\pi$

This question asks for the diameter of the circle. First find the radius, and multiply it by 2:

$$A = \pi r^2$$
$$63\pi = \pi r^2$$
$$63 = r^2$$
$$r^2 = 63$$
$$\sqrt{r^2} = \sqrt{63}$$
$$r = \sqrt{9 \times 7}$$
$$= 3\sqrt{7}$$

The radius is $3\sqrt{7}$ yards. The diameter is twice this length, or $6\sqrt{7}$ yards.

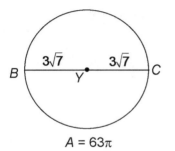

$A = 63\pi$

Degrees in a circle

So far in this chapter, we've learned about the parts of a circle and how to find circumference and area. There are a few other important concepts to know about circles, but before we move on, let's look at degrees.

We saw in the Angles chapter that a straight line measures 180°.

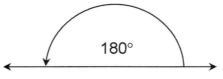

A circle measures 2 times 180, or 360°:

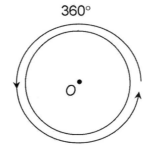

Arcs

A portion of the outer edge of a circle is known as an **arc.** The measure of an entire circle is 360°, but we sometimes need to know the measure of a part of the circle, too.

Arcs are commonly defined by two points on a circle. In this circle, there are two arcs between points A and B.

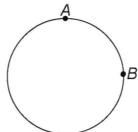

The smaller arc is known as the **minor arc.** It is shown by the arrow in the figure:

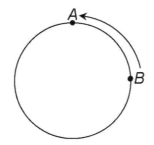

The larger arc is known as the **major arc.** To differentiate between the two arcs, a third point may be added to the circle:

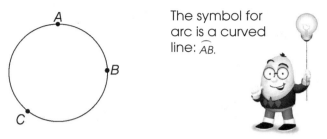

The symbol for arc is a curved line: $\overarc{AB}$.

We refer to the minor arc as arc *AB*. The major arc is labeled with three letters, *ACB*.

Arcs can be measured either in degrees or in units of length, such as inches, feet, or centimeters. When we refer to the **measure** of an arc, we are referring to its number of degrees. When we refer to the **length** of an arc, we are referring to how long it is in terms of inches or some other unit.

Central angles

The degree measure of an arc is equal to the measure of the central angle that intercepts the arc. A **central angle** of a circle is an angle that has its vertex at the origin, or center of the circle. Its sides are formed by two radii.

Examples

Angle *ABC* is a central angle of circle *B*. It measures 90°:

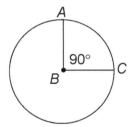

Since m∠*ABC* = 90, this tells us that arc *AC* measures 90°.

In circle *R* below, ∠*QRS* is a central angle that measures 45°. Its vertex is at the center of the circle, point *R*. Its sides are both radii of the circle.

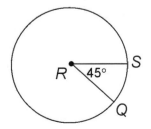

Since ∠*QRS* measures 45°, we know that arc *SQ* measures 45°.

Intercepted arcs

When a central angle intercepts the sides of a circle, it creates two arcs: a minor and a major one. We call the minor arc the **intercepted arc** of the central angle.

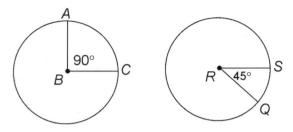

In circle *B* shown, minor arc *AC* is the intercepted arc of central angle *ABC*. Minor arc *SQ* is the intercepted arc of ∠*SRQ* of circle *R*.

Practice questions

1. In the space below, draw circle *L* with central angle *KLM* that measures 60°.

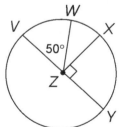

2. In circle *Z* shown, what is the measure of arc *WX*?

3. In circle *D* shown, arc *EF* is the intercepted arc of central angle *EDF*. If arc *EF* measures 135°, what is the value of *x*?

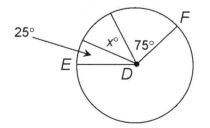

Solutions

1. This circle shows central angle *KLM*, measuring 60°.

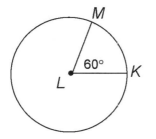

2. The measure of arc *WX* is 40°.

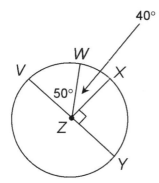

We are told that ∠*XZY* is a right angle. So, we know that ∠*XZV* also measures 90°. The measure of ∠*WZV* is 50°, so ∠*WZX* measures 90 – 50, or 40°. Arc *WX* is the intercepted arc of central angle *WZX*, so arc *WX* also measures 40°.

3. The value of *x* is 35°.

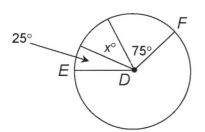

This question gives us an arc measurement and asks for the measure of a central angle. We are told that arc *EF* measures 135°. So, central angle *EDF* also measures 135°. Subtract the two given angles from 135 to find the value of *x*: 135° – 75° – 25° = 35°.

Sectors of circles

A central angle and its intercepted arc together make up what is known as a **sector** of a circle. A sector is a part of the circle in the shape of a wedge.

Examples

In circle *B* shown, sector *ABC* contains central angle *ABC* and intercepted arc *AC*. The central angle measures 90°, and arc *AC* also measures 90°.

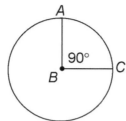

Circle *R* contains sector *SRQ*, with central angle *SRQ* and intercepted arc *SQ*. The measure of ∠*SRQ* is 45°, and the measure of $\widehat{SQ}$ is 45°.

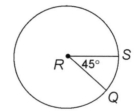

You might be asked to find the area of a sector of a circle. To do this, you must determine what part of the circle the sector represents.

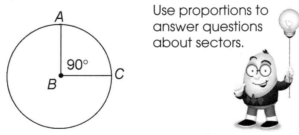

Use proportions to answer questions about sectors.

In circle *B*, sector *ABC* has a central angle of 90°. This central angle is equal to $\frac{90}{360}$ of the entire circle, or $\frac{1}{4}$. Since the central angle represents one-fourth of the whole circle, sector *ABC* also represents one-fourth of the circle.

This relationship can be written as a proportion:

$$\frac{\text{degrees in central angle}}{\text{degrees in entire circle}} = \frac{\text{sector}}{\text{entire circle}}$$

Chapter 7: Circles

We can write it in math terms this way:

$$\frac{90°}{360°} = \frac{\text{sector}}{\text{entire circle}}$$

It can also be reduced, as follows:

$$\frac{1}{4} = \frac{\text{sector}}{\text{entire circle}}$$

You can reduce the fraction or not—it's up to you. However, you may find the reduced fraction easier to work with.

Once we know the portion the sector makes up of the circle, we can find the area of the sector. Let's say the radius of circle *B* is 8:

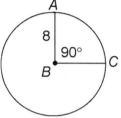

We can find the area of sector *ABC* by finding one-fourth the area of the entire circle. The area of the full circle is 64π square units:

$$A = \pi r^2$$
$$= \pi(8 \times 8)$$
$$= 64\pi$$

Set up a proportion to find the area of the sector:

$$\frac{\text{degrees in central angle}}{\text{degrees in entire circle}} = \frac{\text{area of sector}}{\text{area of entire circle}}$$

Substitute the values you know into the equation:

$$\frac{90°}{360°} = \frac{\text{area of sector}}{64\pi}$$

You might wish to reduce the fraction, as we saw above:

$$\frac{1}{4} = \frac{\text{area of sector}}{64\pi}$$

Now, solve for the area of the sector:

$$\frac{1}{4} = \frac{\text{area of sector}}{64\pi}$$
$$4 \times \text{area of sector} = 64\pi$$
$$\text{area of sector} = \frac{64\pi}{4}$$
$$= 16\pi$$

The area of sector *ABC* is 16π square units.

Practice Questions

1. Circle *R* has a radius of 2 centimeters, as shown in the figure. What is the area of sector *SRQ*?

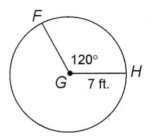

2. Circle *G* has a radius of 7 feet, as shown. Find the area of sector *FGH*.

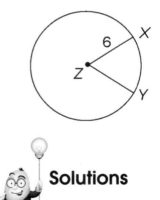

3. In circle *Z* shown, the area of sector *XZY* is 6π square units. What is the measure of central angle *XZY*?

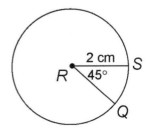

Solutions

1. The area of sector *SRQ* is 0.5π square centimeters.

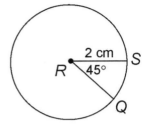

Chapter 7: Circles

Circle *R* has a radius of 2 cm. Find the area of the full circle:

$$A = \pi r^2$$
$$= \pi(2 \times 2)$$
$$= 4\pi$$

The area of the circle is 4π cm^2. Next, find the ratio of the degrees in the central angle to the degrees in the full circle:

$$\frac{\text{degrees in central angle}}{\text{degrees in entire circle}} = \frac{45°}{360°}$$

Set up a proportion to find the area of the sector:

$$\frac{45°}{360°} = \frac{\text{area of sector } SRQ}{\text{area of circle } R}$$
$$\frac{45°}{360°} = \frac{\text{area of sector } SRQ}{4\pi}$$

Cross-multiply to find the area of the sector:

$$\frac{45°}{360°} = \frac{\text{area of sector } SRQ}{4\pi}$$
$$45 \times 4\pi = 360° \times \text{area of sector } SRQ$$
$$180\pi = 360 \times \text{area of sector } SRQ$$
$$\frac{180\pi}{360} = \text{area of sector } SRQ$$
$$0.5\pi = \text{area of sector } SRQ$$

2. The area of sector *FGH* is $16\frac{1}{3}\pi$ ft^2.

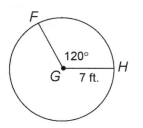

The circle has a radius of 7 feet. So, the area of the full circle is 49π:

$$A = \pi r^2$$
$$= \pi(7 \times 7)$$
$$= 49\pi$$

Next, determine what part the sector represents of the whole circle. The central angle measures 120°, and the full circle is 360°:

$$\frac{\text{degrees in central angle}}{\text{degrees in entire circle}} = \frac{120°}{360°}$$

This can be further reduced to $\frac{1}{3}$. Thus, sector *FGH* is one-third of circle *G*.

Set up a proportion to find the area of the sector:

$$\frac{120°}{360°} = \frac{\text{area of sector } FGH}{\text{area of circle } G}$$

$$\frac{1}{3} = \frac{\text{area of sector } FGH}{49\pi}$$

Cross-multiply to solve:

$$\frac{1}{3} = \frac{\text{area of sector } EFG}{49\pi}$$

$$3 \times \text{area of sector } EFG = 49\pi$$

$$\text{area of sector} = \frac{49\pi}{3}$$

$$= 16\frac{1}{3}\pi$$

The area can also be written as approximately 16.33π square feet.

3. The measure of central angle *XZY* is 60°.

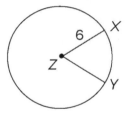

This question gives you the area of the sector, so you have to work backward to find the area of the central angle. Instead of setting up a proportion to find the area of the sector, set up the proportion to find the measure of the angle.

We are told that the area of sector *XZY* is 6π square units. Find the area of the entire circle:

$$A = \pi r^2$$

$$= \pi(6 \times 6)$$

$$= 36\pi$$

The area of the full circle is 36π square units. So, the ratio of the area of the sector to the area of the full circle is

$$\frac{6\pi}{36\pi}, \text{ or } \frac{1}{6}.$$

Set up a proportion to find the measure of the central angle:

$$\frac{1}{6} = \frac{m\angle XZY}{360°}$$

Solve for the measure of $\angle XZY$:

$$\frac{1}{6} = \frac{m\angle XZY}{360°}$$

$$6 \times m\angle XZY = 360°$$

$$m\angle XZY = \frac{360°}{6}$$

$$= 60°$$

Arc length

In addition to finding the area of a sector, you may need to find its perimeter. This requires determining the arc length.

An arc represents a portion of the outer edge of the circle. If we know what portion of the circle the arc represents, we can calculate its length.

Examples

Circle *C* shown below has a radius of 5 meters. Central angle *DCE* intercepts arc *DE* and has a measure of 60°.

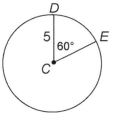

In degrees, the measure of arc *DE* is 60°. But we are looking for its measure in length—specifically, we want to know its number of meters.

To determine the length of the arc, we first determine the circumference of the full circle and the portion of the full circle that the arc represents. Then we can use a proportion.

The circumference of circle *C* is $2\pi r$, or 2 times π times 5, which is 10π meters. The portion that the arc represents of the circle is the same as the ratio of the measure of the central angle to the measure of the full circle:

$$\frac{m\angle DCE}{\text{entire circle}} = \frac{60°}{360°}$$

$$= \frac{1}{6}$$

Arc *DE* represents one-sixth of the circle. Set up a proportion:

$$\frac{1}{6} = \frac{\text{length of } \overset{\frown}{DE}}{\text{circumference}}$$

$$\frac{1}{6} = \frac{\text{length of } \overset{\frown}{DE}}{10\pi}$$

Solve the proportion to find the length of the arc:

$$\frac{1}{6} = \frac{\text{length of } \overset{\frown}{DE}}{10\pi}$$

$$10\pi = 6 \times \text{length of } \overset{\frown}{DE}$$

$$\frac{10\pi}{6} = \text{length of } \overset{\frown}{DE}$$

$$1.67\pi \approx \text{length of } \overset{\frown}{DE}$$

Arc *DE* measures approximately 1.67π meters.

Finding sector perimeters

The perimeter of a sector can be found if we know the radius of the circle plus the length of the sector's arc. In circle *B*, for example, the radius is 8 inches.

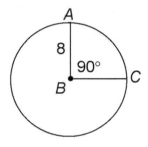

Both sides of the sector are radii of the circle. So, $\overline{BA}$ and $\overline{BC}$ both measure 8 inches. To find the length of $\overset{\frown}{AC}$, first find the circumference of the circle:

$$C = 2\pi r$$
$$= 2\pi(8)$$
$$= 16\pi$$

The circumference is 16π inches. Next, set up a proportion. Start with the ratio of the measure of the central angle to the measure of the full circle:

$$\frac{m\angle ABC}{\text{entire circle}} = \frac{90°}{360°}$$

This ratio is the same as the ratio of the length of arc $\overset{\frown}{AC}$ to the entire circumference. Set up a proportion with the two ratios:

$$\frac{90°}{360°} = \frac{\text{length of } \overset{\frown}{AC}}{\text{circumference}}$$

$$\frac{90°}{360°} = \frac{\text{length of } \overset{\frown}{AC}}{16\pi}$$

Solve the proportion to find the length of the arc:

$$\frac{90°}{360°} = \frac{\text{length of } \overset{\frown}{AC}}{16\pi}$$

$$90 \times 16\pi = 360 \times \text{length of } \overset{\frown}{AC}$$

$$1,440\pi = 360 \times \text{length of } \overset{\frown}{AC}$$

$$\frac{1,440\pi}{360} = \text{length of } \overset{\frown}{AC}$$

$$4\pi = \text{length of } \overset{\frown}{AC}$$

The length of arc AC is 4π inches. The perimeter of the sector is $8 + 8 + 4\pi$, or $16 + 4\pi$ inches.

Practice Questions

1. Circle T has a radius of 3 feet, as shown in the figure. Find the length of arc SU.

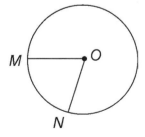

2. In the figure shown, circle O has a radius of 4.5 kilometers. What is the length of arc NP?

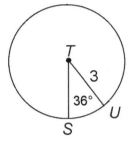

3. Circle O shown has an area of 81π square meters. The area of sector MON is 16.2π square meters. Find the perimeter of sector MON.

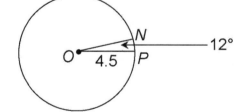

Solutions

1. The length of arc *SU* is 0.6π feet.

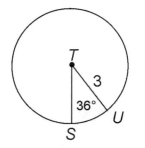

First, find the circumference of circle *T*. The circumference is 2π*r*, or 6π feet.

Next, find the portion of circle *T* that arc SU represents. The measure of central angle *STU* is 36°. This represents $\frac{1}{10}$ of the circle:

$$\frac{m\angle STU}{\text{entire circle}} = \frac{36°}{360°}$$

$$= \frac{1}{10}$$

Now, set up a proportion:

$$\frac{1}{10} = \frac{\text{length of } \overset{\frown}{SU}}{\text{circumference}}$$

$$\frac{1}{10} = \frac{\text{length of } \overset{\frown}{SU}}{6\pi}$$

Cross-multiply to solve:

$$\frac{1}{10} = \frac{\text{length of } \overset{\frown}{SU}}{6\pi}$$

$$6\pi = 10 \times \text{length of } \overset{\frown}{SU}$$

$$\frac{6\pi}{10} = \text{length of } \overset{\frown}{SU}$$

$$0.6\pi = \text{length of } \overset{\frown}{SU}$$

2. The length of arc *NP* is 0.3π kilometers.

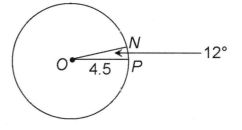

The circumference of circle *O* is 2π*r*, or 9π kilometers. Arc *NP* is intercepted by central angle *NOP*, which measures 12°. Central angle *NOP* represents $\frac{12}{360}$ of the circle, which reduces to $\frac{1}{30}$ of the circle. Set up a proportion and solve:

$$\frac{1}{30} = \frac{\text{length of } \overset{\frown}{NP}}{\text{circumference}}$$

$$\frac{1}{30} = \frac{\text{length of } \overset{\frown}{NP}}{9\pi}$$

$$9\pi = 30 \times \text{length of } \overset{\frown}{NP}$$

$$\frac{9\pi}{30} = \text{length of } \overset{\frown}{NP}$$

$$0.3\pi = \text{length of } \overset{\frown}{NP}$$

3. The perimeter of sector *MON* is 18 + 3.6π meters.

We are looking for the perimeter of sector *MON*. To find the perimeter, we must know the length of the radius and the length of $\overset{\frown}{MN}$.

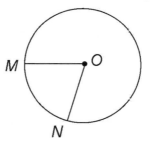

First, find the length of the radius. The area of circle *O* is 81π square meters. Plug this into the area formula and solve for *r*.

$$A = \pi r^2$$

$$81\pi = \pi r^2$$

$$81 = r^2$$

$$\sqrt{r^2} = \sqrt{81}$$

$$r = 9$$

The radius, *r*, measures 9 meters:

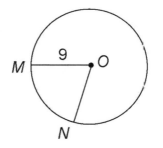

Next, find the length of $\overset{\frown}{MN}$. To do this, we must find the circumference of the circle. We must also know what part of the circumference arc *MN* represents.

Plane Geometry

We are told that the area of circle O is 81π square meters, and the area of sector MON is 16.2π square meters. Set up a ratio with this information:

$$\frac{\text{area of sector } MON}{\text{area of circle } O} = \frac{16.2\pi}{81\pi}$$

$$= \frac{1}{5}$$

This tells us that sector MON is one-fifth of circle O.

Now find the circumference of the whole circle. The radius is 9 meters, so the circumference is $2\pi r$, or 18π meters. Arc MN is one-fifth of that, so the length of $\overset{\frown}{MN}$ is 18π divided by 5, or 3.6π meters.

To find the perimeter of sector MON, add together the lengths of $\overline{OM} + \overline{ON} + \overset{\frown}{MN} = 9 + 9 + 3.6\pi$. The perimeter is $18 + 3.6\pi$ meters.

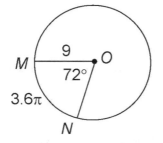

If we had been asked to find the degree measure of arc MN, we could divide the degree measure of the entire circle by 5. The degree measure of the arc is 360° divided by 5, which is 72°.

Inscribed angles

The last type of angle that we will look at in this chapter is the **inscribed angle.** Inscribed angles are angles with a vertex on the outer edge of the circle.

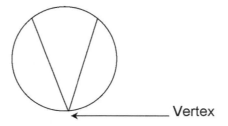

Vertex

They are formed by two chords, which as we saw earlier are lines that extend between any two points on the outer edge of the circle.

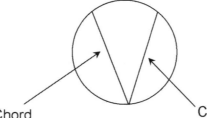

Chord   Chord

Examples

In the circle shown, ∠*QRS* is an inscribed angle measuring 40°. It intercepts arc *QS*. Chords *RS* and *RQ* are the sides of the angle.

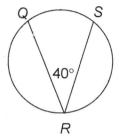

To find the measure of an inscribed angle, use this formula:

$$\text{measure inscribed } \angle = \frac{1}{2} \text{ measure intercepted arc}$$

The intercepted arc of an inscribed angle is always twice the size of the angle itself. In the circle shown, ∠*QRS* is 40°. That tells us that arc *QS* measures 80°.

Practice Questions

1. What is the measure of arc *JL* in the circle shown?

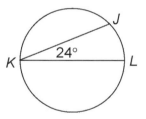

2. In the circle shown, inscribed angle *UVW* measures 55°. Find the measure of arc *UW*.

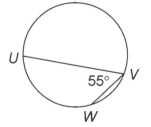

3. If the measure of arc *XZ* is 20° as shown, what is the measure of ∠*XYZ*?

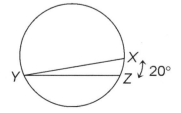

Solutions

1. The measure of arc *JL* is 48°.

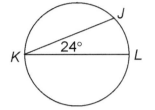

An intercepted arc is twice as large as the measure of its inscribed angle. Angle *JKL* measures 24°, so arc *JL* measures 48°.

2. The measure of arc *UW* is 110°.

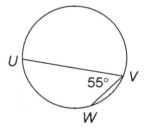

To find the measure of the arc, multiply the measure of the inscribed angle by 2. The arc measures 110°.

3. The measure of ∠*XYZ* is 10°.

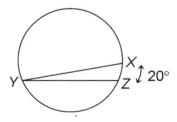

In this question, we are given the measure of arc *XZ* as 20°. Inscribed ∠*XYZ* measures half this amount, or 10°.

Chapter Review

1. The radius *TM* of circle *T* measures 9. What is the length of diameter *MN*?

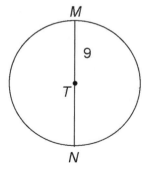

2. In circle *M* shown, what is the length of radius *MO*?

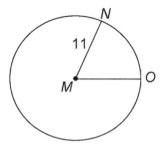

3. Find the circumference of circle *R*.

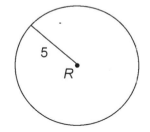

4. The circumference of this circle is 14π. What is the length of its radius?

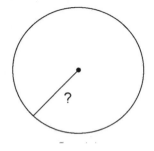

5. What is the area of circle *L*?

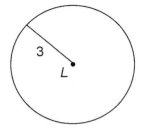

6. What is the radius of circle *T*?

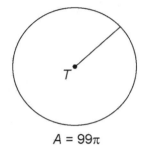

$A = 99\pi$

7. Circle *X* has an area of 2,200π square inches. What is length of $\overline{RS}$?

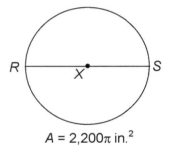

$A = 2{,}200\pi$ in.^{2}

8. In circle *O* shown, what is the measure of ∠*POQ*?

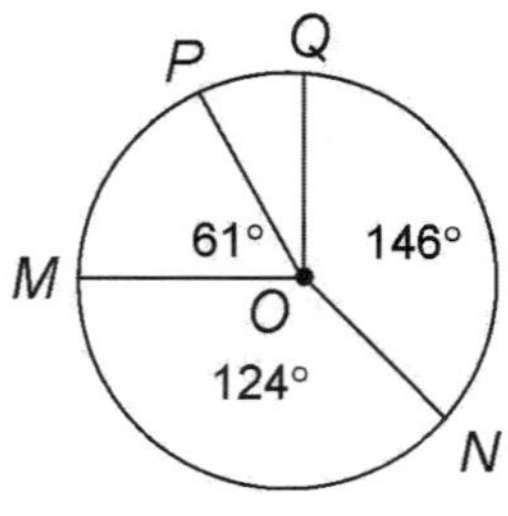

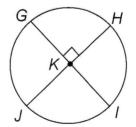

9. In circle *K* shown, what is the measure of arc *HI*?

10. Circle *V* has a radius of 30 units. What is the area of sector *UVW*?

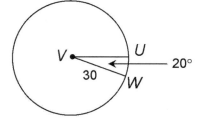

11. In circle *L* shown, the area of sector *KLM* is 12π square inches. What is the measure of central angle *KLM*?

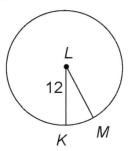

12. Circle *Q* shown has a circumference of 19 centimeters. What is the length, in centimeters, of arc *RP*?

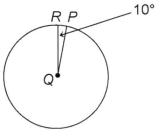

C = 19 cm

13. What is the length of arc *EG* in circle *F* shown?

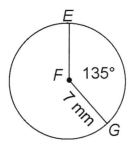

14. In the circle shown, inscribed angle *TSR* measures 15°. What is the measure of arc *TR*?

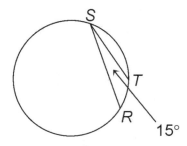

15. Intercepted arc *HJ* measures 130°, as shown in the figure. What is the measure of ∠*HIJ*?

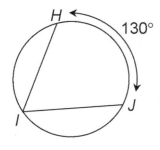

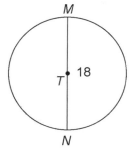 **Solutions**

1. The correct answer is shown below.

The diameter of a circle is twice the length of its radius. To find the length of diameter *MN*, multiply 2 × radius *TM*:

2 × 9 = diameter

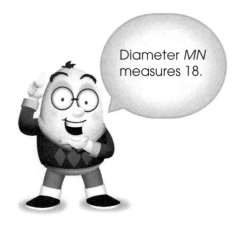

Diameter *MN* measures 18.

2. Radius *MN* measures 11. Every radius on a circle has the same length. So, radius *MO* also measures 11.

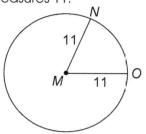

3. To find the circumference, multiply 2 × radius × π. The radius of circle *R* is 5. So, we multiply 2 × 5 × π. The circumference is 10π.

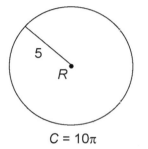

$C = 10\pi$

4. To find the radius, divide the circumference by 2π. The circumference is 14π, so we divide 14π by 2π.

$C = 14\pi$

The radius is 7.

5. To find the area, multiply π × radius × radius. The radius of circle *L* is 3. So, we multiply π × 3 × 3.

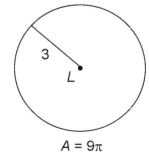

$A = 9\pi$

The area is 9π square units.

6. The radius of circle *T* is $3\sqrt{11}$.

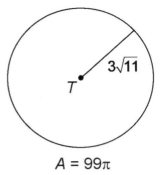

$A = 99\pi$

We are told that the area of the circle is 99π. Substitute this into the area formula:

$$A = \pi r^2$$
$$99\pi = \pi r^2$$
$$99 = r^2$$
$$r^2 = 99$$
$$\sqrt{r^2} = \sqrt{99}$$
$$r = \sqrt{9 \times 11}$$
$$= 3\sqrt{11}$$

7. The length of $\overline{RS}$ is $20\sqrt{22}$ inches.

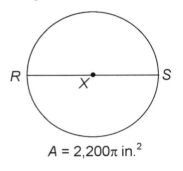

$A = 2,200\pi$ in.^{2}

First, find the length of the radius.

$$A = \pi r^2$$
$$2,200\pi = \pi r^2$$
$$2,200 = r^2$$
$$r^2 = 2,200$$
$$\sqrt{r^2} = \sqrt{2,200}$$
$$r = \sqrt{100 \times 22}$$
$$= \sqrt{10 \times 10 \times 22}$$
$$= 10\sqrt{22}$$

The radius is $10\sqrt{22}$ inches, so the diameter is 2 times $10\sqrt{22}$, or $20\sqrt{22}$ inches.

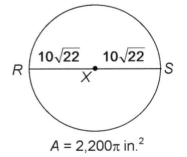

$A = 2,200\pi$ in.^{2}

8. The measure of ∠*POQ* is 29°.

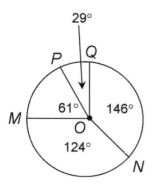

The number of degrees in an entire circle is 360°. The circle shown contains four central angles. The measures of three of the central angles are given; only ∠*POQ* is missing. Subtract the measures of the three given angles from 360°:

360 – 124 – 146 – 61 = 29°.

9. The measure of arc *HI* is 90°.

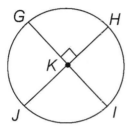

We are shown that circle *K* is divided into four central angles by two diameters, $\overline{HJ}$ and $\overline{GI}$. The diameters are perpendicular, so all of the central angles are right angles. This means each of the four intercepted arcs also measures 90°.

10. The area of sector *UVW* is 50π square units.

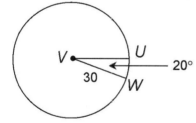

First, determine the ratio of the central angle to the full circle:

$$\frac{\text{degrees in central angle}}{\text{degrees in entire circle}} = \frac{20°}{360°}$$

$$= \frac{1}{18}$$

The sector is therefore $\frac{1}{18}$ of the full circle. Next, determine the area of the full circle. The area of the full circle is 900π square units:

$$A = \pi r^2$$

$$= \pi(30 \times 30)$$

$$= 900\pi$$

Now, set up a proportion to find the area of the sector:

$$\frac{1}{18} = \frac{\text{area of sector}}{\text{area of entire circle}}$$

$$\frac{1}{18} = \frac{\text{area of sector}}{900\pi}$$

Cross-multiply to solve:

$$\frac{1}{18} = \frac{\text{area of sector}}{900\pi}$$

$$18 \times \text{area of sector} = 900\pi$$

$$\text{area of sector} = \frac{900\pi}{18}$$

$$= 50\pi$$

The area of sector *UVW* is 50π. We are not given any units of measurement, so we can just say 50π units^2.

11. The measure of central angle *KLM* is 30°.

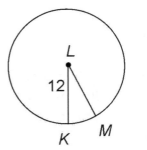

We are told that the area of sector *KLM* is 12π square inches. Using 12 for the radius, we can determine that the area of the circle is 144π square inches:

$$A = \pi r^2$$

$$= \pi(12 \times 12)$$

$$= 144\pi$$

Find the ratio of the area of the sector to the area of the full circle:

$$\frac{\text{area of sector } KLM}{\text{area of circle } L} = \frac{12\pi}{144\pi}$$

$$\frac{\text{area of sector } KLM}{\text{area of circle } L} = \frac{1}{12}$$

Set up a proportion to find the measure of the central angle:

$$\frac{1}{12} = \frac{m\angle KLM}{360°}$$

Solve for the measure of ∠*KLM*:

$$\frac{1}{12} = \frac{m\angle KLM}{360°}$$

$$12 \times m\angle KLM = 360°$$

$$m\angle KLM = \frac{360°}{12}$$

$$= 30°$$

12. The length of arc *RP* is 0.53 centimeters.

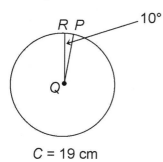

C = 19 cm

To find the length of the arc, first find the portion of the circle that the arc represents. This arc is intercepted by central angle *RQP*, which measures 10°, so the arc also measures 10°. Set up a ratio:

$$\frac{\text{degrees in arc}}{\text{degrees in entire circle}} = \frac{10°}{360°}$$

$$= \frac{1}{36}$$

Arc *RP* represents $\frac{1}{36}$ of circle *Q*.

We are given the circumference of the circle, so we can use the ratio to create a proportion:

$$\frac{1}{36} = \frac{\text{length of arc } RP}{\text{circumference of circle } Q}$$

$$\frac{1}{36} = \frac{\text{length of arc } RP}{19}$$

Cross-multiply to solve:

$$\frac{1}{36} = \frac{\text{length of arc } RP}{19}$$

$$36 \times \text{length of arc } RP = 19$$

$$\text{length of arc } RP = \frac{19}{36}$$

$$\approx 0.53$$

The circumference is given as 19 centimeters, without the symbol π. This means that the value of π has been figured into the circumference. So, we would give the length of the arc as approximately 0.53 centimeters.

13. The length of arc *EG* is 5.25π millimeters.

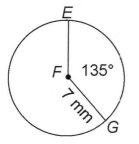

First, find the circumference of circle *F*. The circumference is $2\pi r$, or 14π millimeters. Next, find the portion of the circle that the arc represents. We are told that central angle *EFG* measures 135°:

$$\frac{\text{m}\angle EFG}{\text{entire circle}} = \frac{135°}{360°}$$

Plane Geometry

Use this ratio to create a proportion and solve:

$$\frac{135°}{360°} = \frac{\text{length of } \overset{\frown}{EG}}{\text{circumference}}$$

$$\frac{135°}{360°} = \frac{\text{length of } \overset{\frown}{EG}}{14\pi}$$

$$360 \times \text{length of } \overset{\frown}{EG} = 135 \times 14\pi$$

$$360 \times \text{length of } \overset{\frown}{EG} = 1,890\pi$$

$$\text{length of } \overset{\frown}{EG} = \frac{1,890\pi}{360}$$

$$= 5.25\pi$$

Arc *EG* measures 5.25π millimeters.

14. Arc *TR* measures 30°.

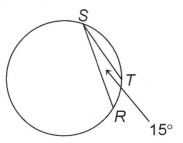

The measure of an inscribed angle is exactly half the measure of its intercepted arc. In this figure, inscribed angle *TSR* measures 15°. Arc *TR* therefore measures 2 × 15, or 30°.

15. Inscribed ∠*HIJ* measures 65°.

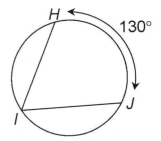

This circle contains an inscribed angle, *HIJ*, with intercepted arc *HJ*. We are given the measure of *HJ*, 130°. To find the measure of ∠*HIJ*, calculate half the measure of its intercepted arc. In this case, the angle measures 130 divided by 2, or 65°.

Chapter 8

Polygons

Hi! I'm egghead. I will teach the following concepts in this chapter:

What is a polygon?

Types of polygons

Naming polygons

Interior angles

Exterior angles

Finding perimeter

Finding area

What is a polygon?

A **polygon** is any closed shape with straight lines for sides. Polygons are two-dimensional, in a single plane.

Types of polygons

Triangles and quadrilaterals are both polygons. A triangle is a three-sided polygon, and a quadrilateral is a polygon with four sides.

There are other types of polygons, and each type has its own name.

A five-sided polygon is called a **pentagon.**

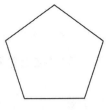

pentagon

A six-sided polygon is a **hexagon.**

hexagon

A seven-sided polygon is a **heptagon.** It can also be called a septagon.

heptagon

These shapes have eight, nine, and ten sides. They are the **octagon, nonagon,** and **decagon.**

octagon   nonagon   decagon

There are others with even more sides! Polygons can be regular or irregular. An **irregular** polygon can have angles of different sizes and sides of varying lengths. The sides and angles of a **regular** polygon, by contrast, are all equal.

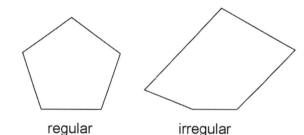

regular   irregular

Most geometry questions you'll see on standardized tests deal with regular polygons.

We'll just work with regular polygons in this chapter.

Naming polygons

Polygons have vertices like other geometry shapes.

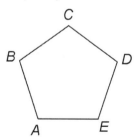

If the vertices of a polygon are labeled, use these letters to refer to the polygon.

Examples

This is pentagon *ABCDE*.

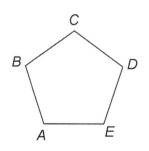

You must always list the letters in vertex order.

This pentagon would **not** be called *ADCEB*.

This is hexagon *UVWXYZ*:

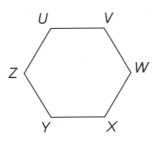

Interior angles

The **interior angles** of polygons are the angles inside the shape.

Just like triangles, other polygons have interior angles as well. These are the interior angles of a pentagon.

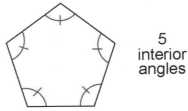

5 interior angles

These are the interior angles of an octagon.

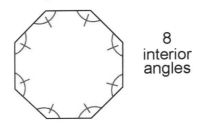

8 interior angles

Regarding interior angles of polygons, there are two types of questions you might see on standardized tests. The first concerns the sum of the interior angles. The second concerns the measure of each individual angle.

To find the sum of the interior angles of a polygon, we use this formula:

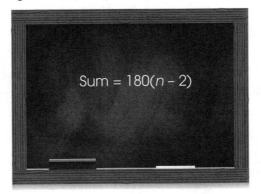

Sum = 180(*n* − 2)

In the formula, *n* stands for the number of sides in the polygon.

Good thing we know the formula!

To find the measure of each interior angle of a polygon, we use this formula:

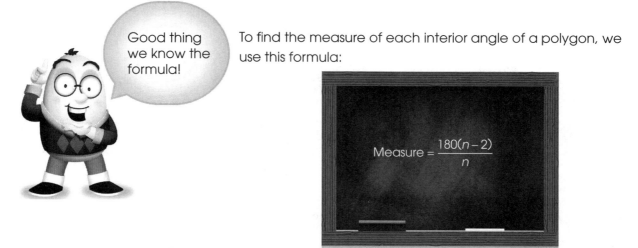

$$\text{Measure} = \frac{180(n-2)}{n}$$

Let's try it with some examples.

Examples

A pentagon has 5 sides. To determine the sum of its interior angles, use the formula $180(n-2)$:

$$\begin{aligned}
\text{Sum} &= 180(n-2) \\
&= 180(5-2) \\
&= 180(3) \\
&= 540
\end{aligned}$$

Here, we substituted 5 for *n*. If we add up the measures of all of the interior angles of a pentagon, the total is 540°.

Next, let's find the measure of each interior angle of a pentagon.

We still use the number 5 for *n*.

Plane Geometry

The formula for the measure of each interior angle is $\dfrac{180(n-2)}{n}$. Substitute 5 for n in the formula:

$$\begin{aligned}
\text{Measure} &= \frac{180(n-2)}{n} \\[6pt]
&= \frac{180(5-2)}{5} \\[6pt]
&= \frac{180(3)}{5} \\[6pt]
&= \frac{540}{5} \\[6pt]
&= 108
\end{aligned}$$

Each interior angle of a pentagon measures 108°.

You will notice that the two formulas are very similar. The measure of each interior angle is found by taking the sum of the interior angles, 180(n – 2), and dividing by the number of sides, n.

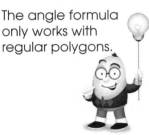

The angle formula only works with regular polygons.

If a polygon is irregular, the formula won't work!

Exterior angles

Along with interior angles, polygons have **exterior angles,** too. The exterior angles are the angles formed outside the shape.

On this pentagon, angle n is an exterior angle.

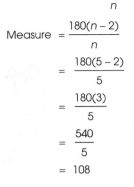

On this octagon, angle x is an exterior angle.

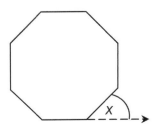

Just like with interior angles, you can find the sum of exterior angles as well as the measure of each exterior angle. You'll see questions about sums and measurements of exterior angles on standardized tests.

The sum of the exterior angles of any polygon is 360°.

To calculate the sum of the exterior angles, we don't have to use a formula. The sum is always the same.

$$\text{Sum} = 360$$

That's pretty simple, right?

To find the measure of each individual exterior angle, divide 360 by the number of sides:

$$\text{Measure} = \frac{360}{n}$$

Here are some examples.

Examples

The sum of the exterior angles of a pentagon is 360°.

The measure of each individual exterior angle is 360 divided by the number of sides:

$$
\begin{aligned}
\text{Measure} &= \frac{360}{n} \\
&= \frac{360}{5} \\
&= 72
\end{aligned}
$$

Each exterior angle measures 72°.

The interior and exterior angles of a polygon are supplementary. They lie on a straight line.

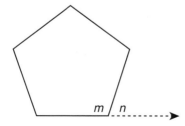

So, we know that the two measures add up to 180°. If we know the measure of the interior angle, we can calculate the measure of the exterior angle, and vice versa.

Practice Questions

1. What is the name of the pentagon in the figure shown?

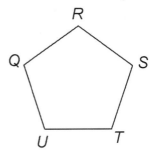

2. Draw a regular heptagon in the space below. Label the figure *JKLMNOP*.

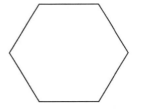

3. What is the sum of the interior angles of the regular polygon shown?

4. What is the measure of each interior angle of the regular polygon shown?

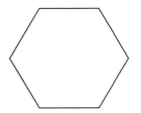

5. What is the measure of each exterior angle of the regular polygon below?

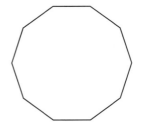

 **Solutions**

1. This is pentagon *QRSTU*.

2. A heptagon is a polygon with 7 sides. Your figure might look like this:

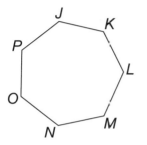

3. The sum of the interior angles is 720°.

Use the formula for the sum of the interior angles of a polygon. This figure is hexagon with 6 sides, so we substitute 6 for *n* in the formula.

$$
\begin{aligned}
\text{Sum} &= 180(n-2) \\
&= 180(6-2) \\
&= 180(4) \\
&= 720
\end{aligned}
$$

4. The measure of each interior angle is 120°.

To determine the measure of each interior angle, use the formula:

$$
\begin{aligned}
\text{Measure} &= \frac{180(n-2)}{n} \\
&= \frac{180(6-2)}{6} \\
&= \frac{180(4)}{6} \\
&= \frac{720}{6} \\
&= 120
\end{aligned}
$$

Again, we substituted 6 here for *n*, the number of sides.

5. Here we have a decagon which has 10 sides. To find the measure of each exterior angle of a regular polygon, we use the formula $\frac{360}{n}$, where *n* is the number of sides.

We substitute 10 for *n* to get:

$$\text{Measure} = \frac{360}{10}$$
$$= 36$$

The measure of each exterior angle is 36°.

Finding perimeter

To find the perimeter of a polygon, we add up the lengths of its sides.

Examples

The regular pentagon below has sides of length 7 centimeters.

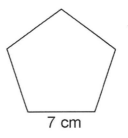

7 cm

The perimeter of the pentagon equals 7 + 7 + 7 + 7 + 7, or 35 centimeters.

Practice Questions

1. What is the perimeter of the regular polygon shown?

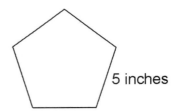

5 inches

2. What is the perimeter of the regular hexagon in the figure shown?

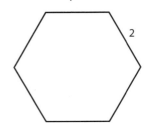

2

3. Pentagon *CDEFG* has a perimeter of 30 feet. What is the length of each side?

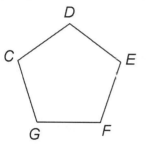

4. Sally works nights as a security guard at a local diamond mine. The grounds of the mine are in the shape of a regular octagon with each side measuring 400 feet, as shown. If Sally walks around the entire grounds exactly one time, how many feet does she walk?

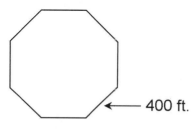

400 ft.

5. The perimeter of the regular polygon below is 49 meters. What is the length of each side?

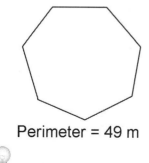

Perimeter = 49 m

 ## Solutions

1. The perimeter of the figure is 25 inches.

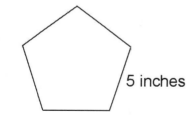

5 inches

The figure is a pentagon with five sides. Each side measures 5 inches. The perimeter is 5 + 5 + 5 + 5 + 5, or 25 inches.

2. The perimeter of the hexagon is 12.

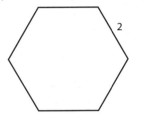

This hexagon has 6 sides, each of length 2. Multiply 2 times 6 to find the perimeter:

$$2 \times 6 = 12$$

3. The length of each side is 6 feet.

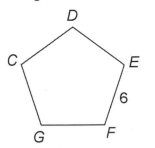

Pentagon *CDEFG* has 5 equal sides. Its perimeter is 30 feet. Divide the perimeter by the number of sides: $30 \div 5 = 6$. Each side measures 6 feet.

4. Sally walks a total of 3,200 feet.

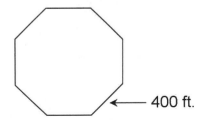

← 400 ft.

The grounds have eight sides measuring 400 feet each. If Sally walks once around the grounds, she walks 8×400, or 3,200 feet.

5. The length of each side is 7 meters.

Perimeter = 49 m

The polygon is a regular heptagon with seven equal sides. To find the length of each side, divide the perimeter, 49, by the number of sides, 7. Each side measures 7 meters.

Finding area

To find the area of a regular polygon with five or more sides, we must have the perimeter of the polygon. We also need another value called the **apothem.**

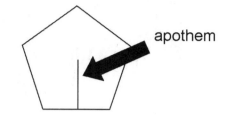

apothem

The apothem of a regular polygon is similar to the radius of a circle. It starts at the center of the shape:

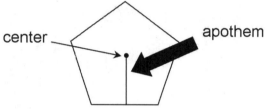

center   apothem

The apothem extends to the outer edge of the polygon. It bisects one of the sides.

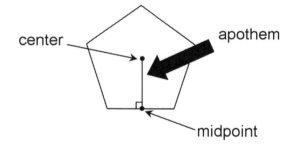

center   apothem

midpoint

To find the area of a regular polygon, we take half of the apothem times the perimeter. Here is the formula:

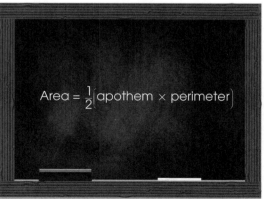

$$\text{Area} = \frac{1}{2}\left(\text{apothem} \times \text{perimeter}\right)$$

Plane Geometry

In shorthand, the formula looks like this:

$$A = \frac{1}{2}(a \times p)$$

It can also be written like this:

$$A = \frac{a \times p}{2}$$

Examples

The pentagon below has a perimeter of 30 and an apothem of 4.13.

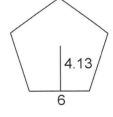

The area of the pentagon equals $\frac{1}{2}a \times p$:

$$A = \frac{1}{2}a \times p$$
$$= \frac{1}{2}(4.13 \times 30)$$
$$= \frac{1}{2}(123.90)$$
$$= 61.95$$

The area of the pentagon equals 61.95 square units.

Practice Questions

1. What is the area of the figure shown, to the nearest two decimal places?

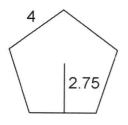

2. What is the area of the figure shown, to the nearest two decimal places?

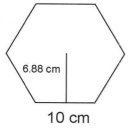

3. Sally the patrolwoman walks around the grounds of a diamond mine, as shown in the figure. What is the area of the mining property that Sally patrols, to the nearest two decimal places?

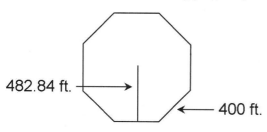

4. John and David have a six-sided table, as shown in the figure. What is the area of the table, to the nearest two decimal places?

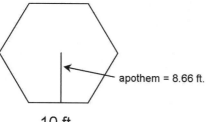

Solutions

1. The area is $\frac{1}{2}(2.75 \times 20)$, or 27.50 square units.

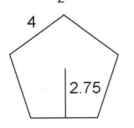

The pentagon has sides of length 4 and an apothem of length 2. Its perimeter is 4 × 5, or 20.

Use the area formula:

$$A = \frac{1}{2}a \times p$$
$$= \frac{1}{2}(2.75 \times 20)$$
$$= \frac{1}{2}(55)$$
$$= 27.50$$

The area of the pentagon equals 27.50 square units.

2. The area is $\frac{1}{2}(6.88 \times 60)$, or 206.40 square centimeters.

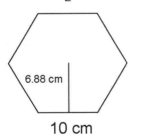

The hexagon shown has sides of 10 centimeters and an apothem of 6.88 centimeters. Since the sides measure 10 centimeters, its perimeter equals 60 centimeters.

The area of the hexagon equals $\frac{1}{2}a \times p$:

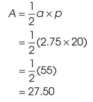

$$A = \frac{1}{2}a \times p$$
$$= \frac{1}{2}(6.88 \times 60)$$
$$= \frac{1}{2}(412.80)$$
$$= 206.40$$

3. The area of the mining property is 772,544 square feet.

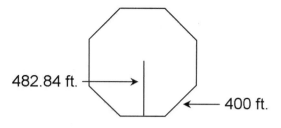

To find the area, multiply $\frac{1}{2}a \times p$. In this case, the apothem is 482.84 feet, and the perimeter is

400×8, or 3,200 feet. The area equals $\frac{1}{2}(482.84 \times 3,200)$, or 772,544 square feet.

4. The area of the table is 259.80 square feet.

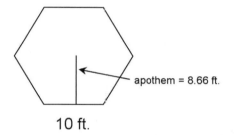

To find the area of the table, use the area formula. The apothem is 8.66 feet, and the perimeter is 10×6, or 60 feet.

$$A = \frac{1}{2}a \times p$$
$$= \frac{1}{2}(8.66 \times 60)$$
$$= \frac{1}{2}(519.60)$$
$$= 259.80$$

The area of the table is 259.80 square feet.

Chapter Review

1. What is the name of the octagon in the figure shown?

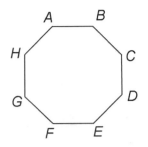

2. Draw a regular decagon with sides of length 10 in the space below.

3. What is the sum of the interior angles of the regular figure shown?

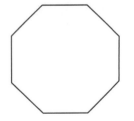

4. What is the measure of each interior angle of the regular figure shown?

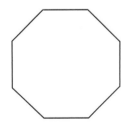

5. What is the sum of the exterior angles of the regular figure shown?

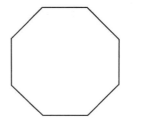

6. What is the measure of each exterior angle of the regular figure shown?

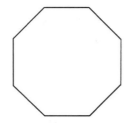

7. What is the perimeter of the regular figure shown?

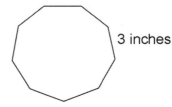 3 inches

8. What is the perimeter of the regular decagon in the figure shown?

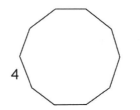 4

9. The regular octagon shown has a perimeter of 64 centimeters. What is the length of each side?

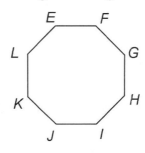

10. Angel creates a garden in the shape of a hexagon. The garden is lined with six hedges around its outer edge, as shown. If each of the hedges has length 12.5 feet, what is the perimeter of the garden?

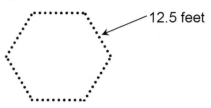

12.5 feet

11. Three students cut a piece of fabric in the shape of a regular pentagon for an art project. The pentagon has a perimeter of 17 feet. What is the measure, in feet, of side x shown?

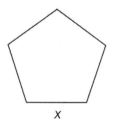

x

12. What is the area of the regular figure shown, to the nearest two decimal places?

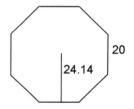

20

24.14

13. What is the area of the regular figure shown, to the nearest two decimal places?

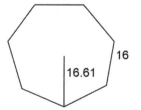

16

16.61

14. Angel's garden from Question 10 above has sides of length 12.5 feet. The distance from the center of the garden to the midpoint of one hedge is exactly 10.83 feet, as shown. What is the area, in square feet, of the garden, to the nearest two decimal places?

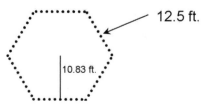

12.5 ft.

10.83 ft.

15. The pentagonal piece of fabric from Question 11 has a perimeter of 17 feet. Its apothem is 2.34 feet. What is the area, in square feet, of the fabric, to the nearest two decimal places?

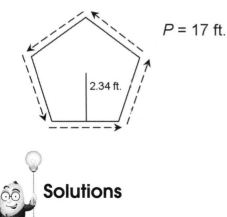

$P = 17$ ft.

2.34 ft.

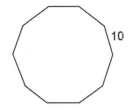

Solutions

1. This is octagon *ABCDEFGH*.

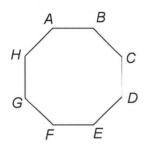

2. Here is what a regular decagon with sides of length 10 might look like.

10

3. The sum of the interior angles is 1,080°.

This figure is an octagon with 8 sides. Substitute 8 for *n* in the formula:

$$
\begin{aligned}
\text{Sum} &= 180(n - 2) \\
&= 180(8 - 2) \\
&= 180(6) \\
&= 1{,}080
\end{aligned}
$$

4. The measure of each interior angle is 135°.

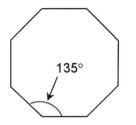

To determine the measure of each interior angle, use the formula:

$$\text{Measure} = \frac{180(n-2)}{n}$$

$$= \frac{180(8-2)}{8}$$

$$= \frac{180(6)}{8}$$

$$= \frac{1{,}080}{8}$$

$$= 135$$

Since the figure is an octagon, we substituted 8 here for n.

5. The sum of the exterior angles of an octagon is 360°.

6. The measure of each exterior angle of the octagon is 45°.

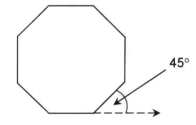

Use the formula $\frac{360}{n}$. Substitute 8 for n:

$$\text{Measure} = \frac{360}{n}$$

$$= \frac{360}{8}$$

$$= 45$$

7. The perimeter of the figure is 27 inches.

3 inches

This shape is a nonagon with 9 equal sides. Each side measures 3 inches. The perimeter is 3 × 9, or 27 inches.

8. The perimeter of the decagon is 40.

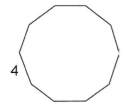

4

The sides of this decagon each measure 4. There are 10 sides, so we can multiply to find the perimeter:

$$4 \times 10 = 40$$

9. The length of each side is 8 centimeters.

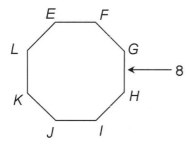

E   F
L   G
K   H
J   I
← 8

The octagon has 8 equal sides. We are given its perimeter, 64 centimeters. Divide the perimeter by the number of sides: 64 ÷ 8 = 8 centimeters per side.

10. The perimeter of the garden is 75 feet.

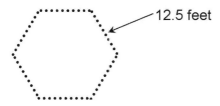

12.5 feet

The garden has 6 equal sides, each measuring 12.5 feet. To find the perimeter, we multiply 6 by 12.5, which equals 75 feet.

11. The measure of side *x* is 3.4 feet.

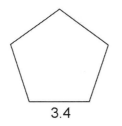

3.4

The pentagon has 5 equal sides and a perimeter of 17 feet. To find the length of each side, divide 17 by 5. The measure of *x* is 3.4 feet.

12. The area is $\frac{1}{2}(24.14 \times 160)$, or 1,931.20 square feet.

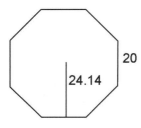

20

24.14

The octagon has sides of length 20 and an apothem of length 24.14. Its perimeter is 8 × 20, or 160.

Use the area formula:

$$A = \frac{1}{2}a \times p$$
$$= \frac{1}{2}(24.14 \times 160)$$
$$= \frac{1}{2}(3,862.40)$$
$$= 1,931.20$$

The area of the octagon is 1,931.20 square feet.

13. The area is $\frac{1}{2}(16.61 \times 112)$, or 930.16 square feet.

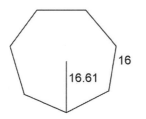

16

16.61

This figure is a heptagon with 7 equal sides. The sides measure 16, and the apothem measures 16.61. To find the perimeter, multiply 7 by 16. The perimeter is 112.

Use the formula for the area of a polygon, $\frac{1}{2}a \times p$:

$$A = \frac{1}{2}a \times p$$
$$= \frac{1}{2}(16.61 \times 112)$$
$$= \frac{1}{2}(1,860.32)$$
$$= 930.16$$

14. The area of Angel's garden is 406.13 square feet.

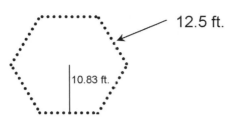

12.5 ft.

10.83 ft.

The perimeter of the garden is 12.5 × 6, or 75 feet. Substitute 10.83 for a and 75 for p into the formula:

$$A = \frac{1}{2}a \times p$$
$$= \frac{1}{2}(10.83 \times 75)$$
$$= \frac{1}{2}(812.25)$$
$$= 406.13$$

The area of the garden is 406.13 square feet.

15. The area of the piece of fabric is 19.89 square feet.

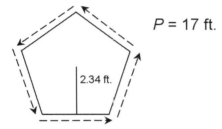

$P = 17$ ft.

2.34 ft.

The perimeter is 17 feet, and the apothem is 2.34 feet. Using the area formula, we multiply $\frac{1}{2}a \times p$:

$$A = \frac{1}{2}a \times p$$
$$= \frac{1}{2}(2.34 \times 17)$$
$$= \frac{1}{2}(39.78)$$
$$= 19.89$$

We can abbreviate the answer as 19.89 ft².

Chapter 9

Irregular and Multiple Figures

Hi! I'm egghead. I will teach the following concepts in this chapter:

Irregular figures
Finding perimeter
Finding area
Rules of diagonals
Multiple figures
Inscribed figures

Irregular figures

Irregular figures in geometry are those that do not have predictable properties like squares, rectangles, and other regular polygons do.

Here are some examples:

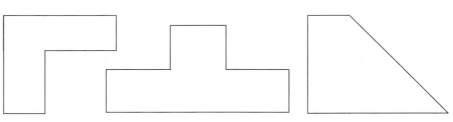

Finding perimeter

Finding the perimeter of irregular figures usually requires finding the lengths of some missing sides. To do this, it's easier if you break the figures up into regular shapes.

Examples

To find the perimeter of this figure, we just need to find the lengths of the missing sides. All of the angles are right angles.

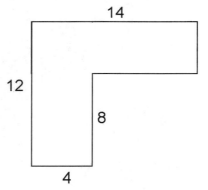

Use subtraction to determine the missing lengths: 12 – 8 = 4, and 14 – 4 = 10.

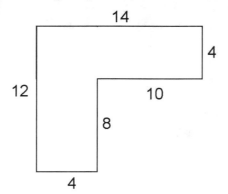

Next, add everything up. The perimeter is 12 + 4 + 8 + 10 + 4 + 14, which is 52.

Let's try another example. This irregular figure has three missing sides. All of the angles are right angles.

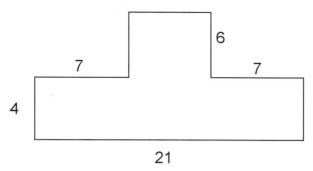

We can find the missing lengths using subtraction: 21 – 7 – 7 = 7. The other two missing sides are congruent to the sides labeled 4 and 6.

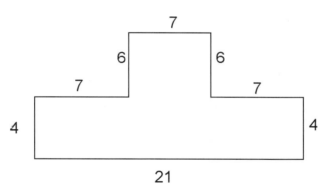

Next, add up the lengths of the sides:

$$21 + 4 + 7 + 6 + 7 + 6 + 7 + 4 = 62$$

The perimeter of the figure is 62.

Practice Questions

1. What is the perimeter of the figure shown?

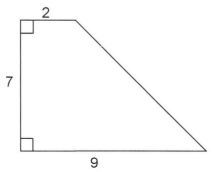

2. Find the perimeter of the irregular quadrilateral shown.

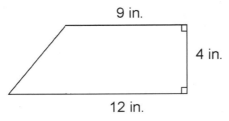

9 in.

4 in.

12 in.

3. A concrete skate park in the shape of a semi-circle has a diameter of 24 meters. A skater skates all the way around the outer edge of the park. How many meters does the skater travel?

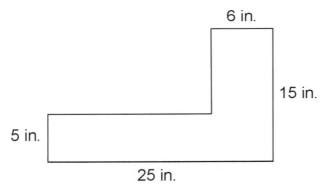

24 meters

4. A building contractor gets a new job to finish a staircase. He has to figure out the perimeter of the staircase already built to build a replica. All intersecting lines meet at right angles, as shown. What's the perimeter of the staircase?

6 in.

15 in.

5 in.

25 in.

5. Find the perimeter of the figure shown.

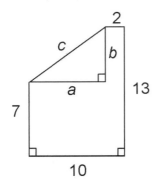

2

c

b

a

13

7

10

Solutions

1. The perimeter of the figure is $18 + 7\sqrt{2}$.

Divide the figure into a rectangle and a triangle. The rectangle has a length of 7 and a width of 2. The triangle has two legs that measure 7. So, the two angles opposite these legs are equal, and they measure 45°.

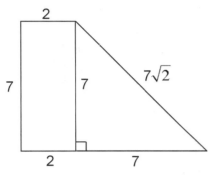

This is a 45-45-90 triangle, with side lengths in a ratio of $1:1:\sqrt{2}$. If two legs measure 7, the hypotenuse must measure $7\sqrt{2}$.

Add up the lengths of the sides to find the perimeter: $7 + 2 + 7 + 2 + 7\sqrt{2} = 18 + 7\sqrt{2}$.

2. The perimeter of the figure is 30 inches.

Divide the figure into regular shapes that you can work with easily. If we draw a perpendicular line as shown, the figure divides into a triangle and a rectangle.

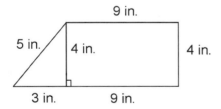

The perpendicular line has length 4 inches. The bottom of the figure separates into 9 inches and 3 inches, as shown. This shows us we have a 3-4-5 right triangle. The hypotenuse measures 5 inches.

Adding up the lengths of the sides, we find the perimeter measures 9 + 4 + 9 + 3 + 5, or 30 inches.

3. The skater travels a total of 12π + 24 meters.

The diameter of the semi-circle is 24 meters. The radius measures half this length, or 12 meters.

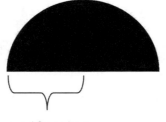

r = 12 meters

To find the circumference of the semi-circle, we calculate half the circumference of the full circle. A circle with radius 12 meters has a circumference of 2 times π times 12, or 24π meters. We want the circumference of the semi-circle, so we take half of 24 meters. The circumference of the semi-circle is 12π meters.

The skater travels not just along the curved edge of the semi-circle, but along the straight edge too. The total distance traveled is 12π + 24 meters.

4.

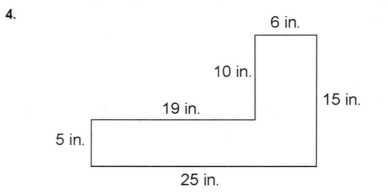

First, determine the length of the missing sides. Then add the side lengths to obtain the perimeter.

To find the lengths of the missing sides, use subtraction. For the horizontal side, subtract 25 – 6. The missing horizontal side measures 19 inches. For the vertical side, subtract 15 – 5. The missing vertical side measures 10 inches.

Now add up the lengths of the sides: 25 + 5 + 19 + 10 + 6 + 15 = 80. The perimeter of the figure is 80 inches.

5. The figure contains a right triangle with sides labeled a, b, and c. To find the perimeter of the shape, we must determine the length of c.

We can find the lengths of a and b using subtraction. The horizontal leg, a, measures 10 – 2 = 8 inches. The vertical leg, b, measures 13 – 7 = 6 inches. Mark these on the figure as shown:

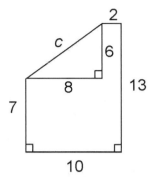

The right triangle has two legs measuring 6 and 8, so it is a 6-8-10 Pythagorean triple. The hypotenuse, c, measures 10.

Add up the lengths of the sides of the figure: 10 + 2 + 13 + 10 + 7 = 42. The perimeter is 42 units.

Finding area

To find the area of irregular figures, again it helps to divide the figure into regular shapes. We find the area of each shape, and add the areas together.

Examples

Here is a figure that we saw in the section on perimeters above. All lines meet at right angles.

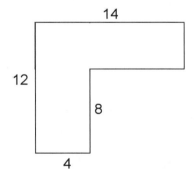

Plane Geometry

We can find the area by dividing the figure into two rectangles. The upper rectangle has an area of 14 × 4 = 56 square units. The lower rectangle has an area of 8 × 4 = 32 square units. The area of the figure is 56 + 32 = 88 units^2.

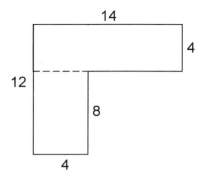

Here is a second example. Again in this figure, all angles are right angles.

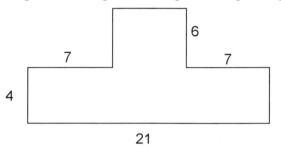

This figure can also be divided into two rectangles.

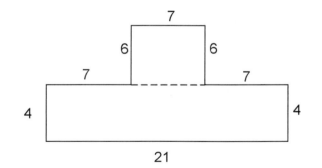

We can then find the area of the rectangles and add them up. The top rectangle has an area of 7 × 6 = 42 units^2. The bottom rectangle has an area of 4 × 21 = 84 units^2. The total area is 126 square units.

Practice Questions

1. What is the area of the figure shown?

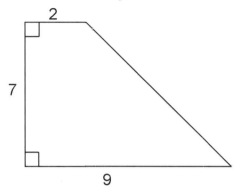

2. Find the area of the irregular quadrilateral shown.

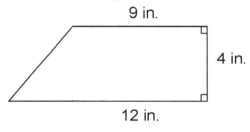

3. What is the area of the concrete skate park shown in the figure?

24 meters

4. Find the area of the figure below.

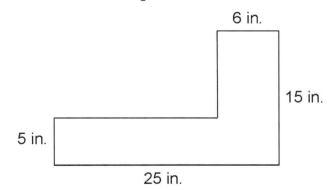

5. Find the area of the figure below.

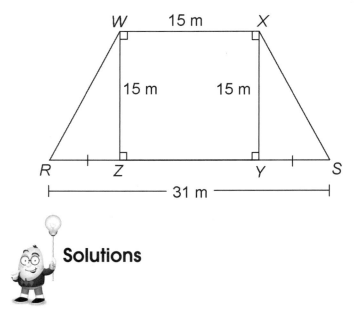

 **Solutions**

1. The area of the figure is 38.50 square units.

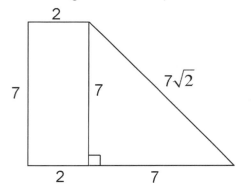

By dividing the figure into a rectangle and a triangle, we can find the lengths of the missing sides. The rectangle has an area of 7 times 2, or 14 square units. The triangle has an area of $\frac{1}{2}$ times 7 times 7, or 24.50 square units. Added together, these areas give us a total of 38.50 units^2.

2. The area of the figure is 42 in^2.

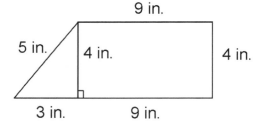

The figure divides into a triangle and a rectangle. The rectangle has an area of 4 times 9, or 36 square inches. The triangle has an area of $\frac{1}{2}$ times 3 times 4, or 6 square inches. The total area is $36 + 6 = 42$ in^2.

Chapter 9: Irregular and Multiple Figures

3. The area of the skate park is 72π square meters.

r = 12 meters

The skate park has a diameter of 24 meters and a radius of 12 meters. To find the area of a circle, we calculate π times *r* times *r*, which in this case is π times 12 times 12, or 144π.

Don't stop there, however! The question asks for the area of the semi-circle, which is half of that value. The area of the park is 72π m².

4. The area is 185 square inches.

First, find the lengths of the missing sides. The missing horizontal side is 25 – 6, or 19 inches. The missing vertical size measures 15 – 5 = 10 inches.

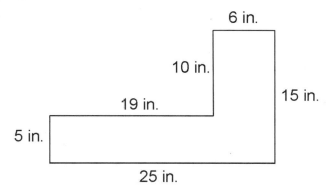

Next, divide the figure into two rectangles. Two possible rectangles are shown by the dotted line:

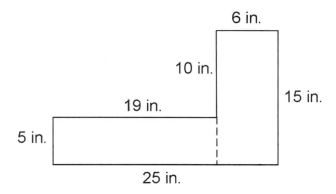

One rectangle measures 19 by 5 inches, and the other measures 6 by 15 inches. Calculate the area of both rectangles: 19 × 5 = 95 square inches, and 6 × 15 = 90 square inches. Added together, this gives a total area of 95 + 90 = 185 square inches.

5. The area of the figure is 345 square meters.

This figure contains a square, *WXYZ*, and two right triangles.

To find the total area, we must determine the area of the square plus the area of the two triangles. The area of the square equals 15 × 15, or 225 square inches.

To find the area of the triangles, we must find the lengths of bases $\overline{RZ}$ and $\overline{YS}$. The length of line segment *RS* is given as 31 meters. Subtract the length of side $\overline{ZY}$ from side $\overline{RS}$: 31 – 15 = 16 meters. Since $\overline{RZ}$ and $\overline{YS}$ are congruent, we can divide 16 by 2 to get the measure of each: $\overline{RZ}$ and $\overline{YS}$ each measure 8 meters.

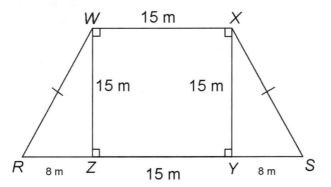

Now, determine the area of each triangle:

$$A = \frac{1}{2}(b \times h)$$
$$= \frac{1}{2}(8 \times 15)$$
$$= \frac{1}{2}(120)$$
$$= 60$$

The area of each triangle is 60 square meters. Add together the three areas: 60 + 225 + 60 = 345. The area of the figure is 345 square meters.

Rules of diagonals

Before we move on to multiple figures, there are some properties of diagonals that are helpful to know. A **diagonal** is a straight line drawn from one corner of a polygon to another corner that is not adjacent.

Here is one diagonal of a rectangle:

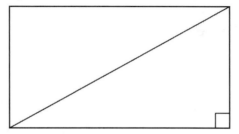

Diagonals commonly appear on geometry test problems, particularly those problems that involve quadrilaterals. The diagonals of each shape have different properties.

Parallelograms

The diagonals of a parallelogram bisect each other.

Each diagonal is divided into two equal parts, but the diagonals are **not** congruent.

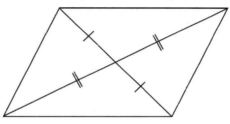

Rectangles

The diagonals of a rectangle bisect each other. They are also congruent.

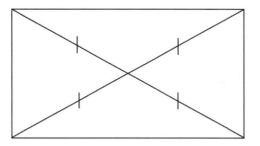

Rhombuses

The diagonals of a rhombus bisect each other. They are also perpendicular.

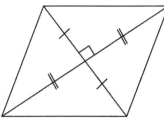

They also bisect the interior angles of the rhombus:

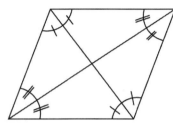

However, they are **not** congruent.

Parallelograms and rhombuses are slanted, so one diagonal is longer.

Squares

Squares have the properties of both rhombuses and rectangles.

The diagonals of a square bisect each other and are congruent. They are perpendicular, and they bisect the interior angles of the square.

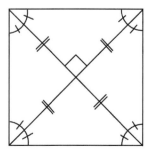

In fact, because we know that squares have four right angles, we can tell the exact measurement of the angles that are formed when the diagonals are drawn in:

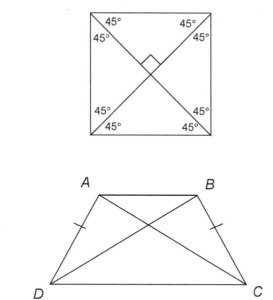

Trapezoids

Trapezoids don't have many diagonal rules. For an isosceles trapezoid, diagonals are congruent. No other types of trapezoids have special diagonal properties.

Multiple figures

Along with problems involving irregular figures, some geometry exams contain questions about multiple figures. Like with irregular figures, multiple figures can be highly varied.

Two triangles

One common type of multiple figures question concerns two triangles that are connected in some way. The triangles may be drawn inside a figure or joined by shared angles or sides.

Examples

Here are some examples. This figure highlights two sets of two triangles created when diagonals are drawn in a rhombus:

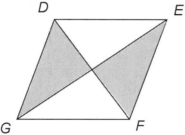

Plane Geometry

This figure has two triangles joined at a single point:

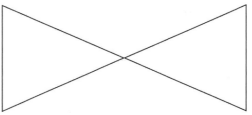

Two triangles can also be created by a single diagonal drawn in a rectangle or other quadrilateral:

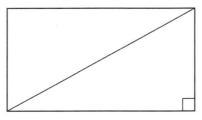

With questions involving two triangles, it's always helpful to look for the special right triangles, such as 30-60-90 triangles, 45-45-90 triangles, and Pythagorean triples.

Inscribed figures

Geometry questions also contain figures that are drawn inside other figures. We call these **inscribed figures.**

Examples

Inscribed figures can have a great degree of variation. Here is a triangle drawn within a square:

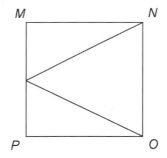

We might also have a quadrilateral within a triangle:

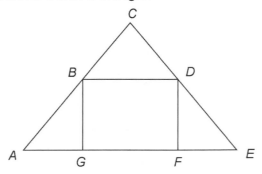

Chapter 9: Irregular and Multiple Figures

Other possibilities are triangles within circles, circles inside triangles, circles inside squares, and squares inside circles:

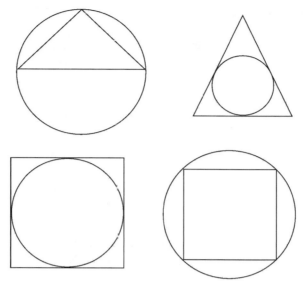

There are many possible multiple-figure combinations. To answer questions about them, draw on the properties of the shapes you know.

In the figure shown, *BDFG* is a rectangle and $\triangle CBD$ is equilateral. What is the measure of $\angle BAG$?

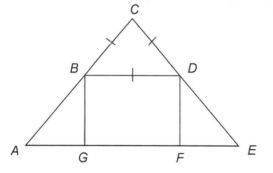

The measure of $\angle BAG$ can be found by applying the properties of certain triangles. First, equilateral triangles have three equal angles and three equal sides. All angles measure 60°. Draw these into the figure, as shown.

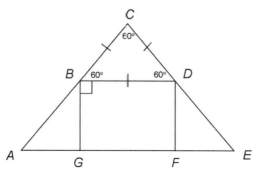

Plane Geometry

Next, we can find the value of ∠ABG by using the properties of straight lines. Every straight line measures 180°. The angles that form a straight line are supplementary. So, ∠CBD, ∠DBG, and ∠ABG are supplementary. This can help us find the value of ∠ABG:

$$m\angle CBD + m\angle DBG + m\angle ABG = 180°$$
$$60 + 90 + m\angle ABG = 180°$$
$$150 + m\angle ABG = 180°$$
$$m\angle ABG = 180° - 150°$$
$$m\angle ABG = 30°$$

Knowing that ∠ABG measures 30°, we can determine the value of ∠BAG: 180 − 90 − 30 = 60°.

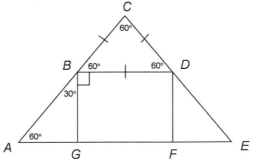

Practice Questions

1. If DF = 6 and GE = 8, what is the perimeter of the rhombus shown?

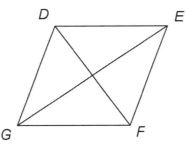

2. Triangle XZY is inscribed in a circle with center O. If the base of the triangle $\overline{XY}$ measures 4 yards, what is the radius of circle O?

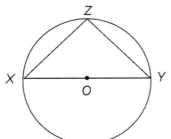

3. A circle with center *Z* is inscribed in a square, as shown. If the sides of the square each measure 10 inches, what is the area of circle *Z*?

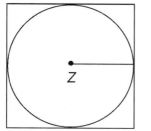

4. The height of the triangle shown in the figure is 14. What is the area of the shaded portion of the figure, to the nearest two decimal places?

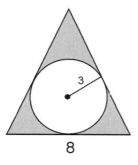

5. The rhombus shown has side lengths measuring 12 inches. It is divided into four triangles by its two diagonals. If the longest diagonal measures 18 inches, what is the area of one of the triangles?

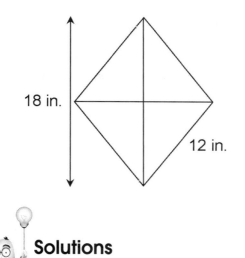

Solutions

1. The perimeter of the rhombus is 20.

This rhombus has one diagonal measuring 6 and another measuring 8. The diagonals of a rhombus bisect each other at right angles. They form four right triangles, as shown:

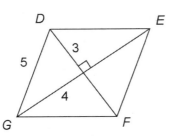

Since *DF* = 6, the leg of one triangle is 3. And since *GE* = 8, the other leg of the triangle is 4. This is a 3-4-5 triangle, with a hypotenuse measuring 5. The hypotenuse is the side of the rhombus, so the perimeter of the rhombus is 5 + 5 + 5 + 5, or 20.

2. The radius of circle *O* is 2 yards.

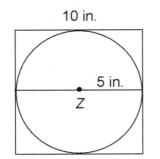

Here we are told that *XY* = 4 yards. The base of the triangle runs through the center of the circle, *O*. So, the base of the triangle is a diameter of the circle. The radius equals half the length of the diameter. This means the radius measures 2 yards.

3. The area of circle *Z* is 25π square inches.

We are told that the sides of the square measure 10 inches. The diameter of circle *Z* is the same length as the side of the square. The diameter of circle *Z* therefore measures 10 inches, and the radius measures 5.

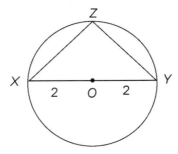

Use the formula for the area of a circle:

$$A = \pi r^2$$
$$= \pi 5^2$$
$$= \pi \times 5 \times 5$$
$$= 25\pi$$

The area of the circle is 25π in^2.

Chapter 9: Irregular and Multiple Figures

4. The area of the shaded portion of the figure is approximately 27.74 square units.

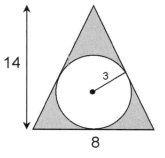

Subtract the area of the circle from the area of the full triangle. This will leave us with the area of the shaded portion of the figure.

The base of the triangle is 8 units, and the height is 14. The area of the entire triangle is $\frac{1}{2}(8 \times 14)$, or 56 square units. The area of the circle is $\pi \times 3^2$, or 9π square units. The area of the shaded portion is $56 - 9\pi$, or approximately 27.74 square units.

5. The area of one triangle is $13.50\sqrt{7}$ square inches.

To find the area of one triangle, we must know its base and height. According to the rules of diagonals, the diagonals of a rhombus bisect each other. The longest diagonal is 18 inches, so half of this is 9 inches. We can mark this on the figure as the height of one of the triangles. We have marked the base of the triangle b:

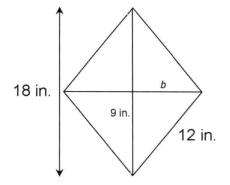

To find the length of b, use the Pythagorean theorem:

$$9^2 + b^2 = 12^2$$
$$81 + b^2 = 144$$
$$b^2 = 144 - 81$$
$$b^2 = 63$$
$$b = \sqrt{63}$$
$$b = \sqrt{9 \times 7}$$
$$b = 3\sqrt{7}$$

The length of b is $3\sqrt{7}$ inches.

Substitute the values of b and h into the formula for the area of a triangle:

$$A = \frac{1}{2}(b \times h)$$
$$= \frac{1}{2}\left(3\sqrt{7} \times 9\right)$$
$$= \frac{1}{2}\left(27\sqrt{7}\right)$$
$$= 13.50\sqrt{7}$$

The area of one triangle is $13.50\sqrt{7}$ square inches.

Chapter Review

1. In the figure below, quadrilateral *MOPQ* is a square. What is the perimeter of *MNOPQ*?

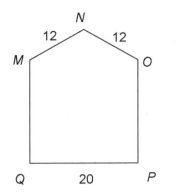

2. In the figure below, quadrilateral *MOPQ* is a square. What is the area, in square units, of pentagon *MNOPQ*?

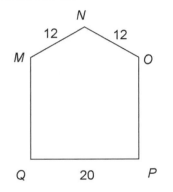

3. What is the area, in square units, of the figure shown? All angles in the figure are right angles.

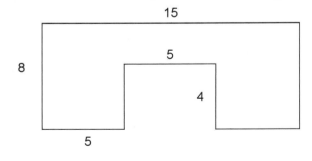

Chapter 9: Irregular and Multiple Figures

4. The map below shows a downtown shopping district. Six stores in the district are labeled *B, C, D, E, F,* and *G*. The numbers on the figure represent the number of blocks between each store. All of the corners are right angles.

Bonnie starts at store *B* and walks to store *E*. Assuming she must use one of the routes shown in the figure below, what is the minimum number of blocks Bonnie must walk to reach store *E*?

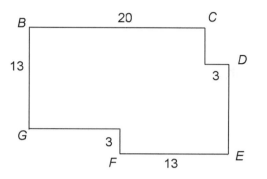

5. Rectangle *EFGH* has a length of 15 meters and a width of 8 meters, as shown in the figure. What is the perimeter of △*OHG*?

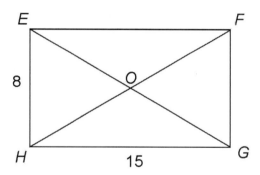

6. Rectangle *PQRS* has diagonals that intersect at point *T*, as shown. What is the measure of *x*?

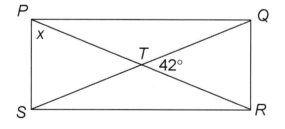

Plane Geometry

7. In the figure shown, what is the value of *a*?

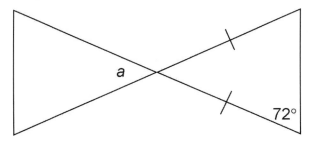

8. If $AD = 6$ inches, what is the length of diagonal $\overline{BD}$ in the rectangle shown?

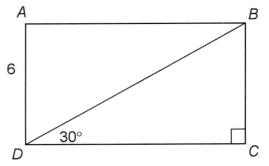

9. A rectangular swimming pool is surrounded by a brick walkway that measures 2 feet in width. The outer dimensions of the walkway are 40 feet by 52 feet as shown. What is the area of the swimming pool?

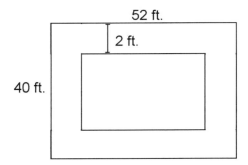

10. A rhombus is inscribed in rectangle *HIJK*, as shown in the figure. What is the measure of $\overline{IJ}$?

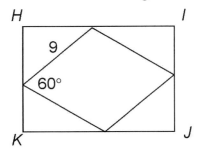

11. A circle with center R is inscribed in a triangle, as shown. Line segment $\overline{RO}$ bisects the base of the triangle at point O. If $PN = 8$ feet, and $RN = 5$ feet, what is the area of circle R?

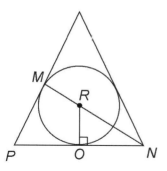

12. Right triangle RST is inscribed in circle T as shown. What is the value of z?

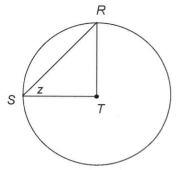

13. A square is inscribed in a circle as shown. If the sides of the square measure 12 units, what is the measure of the diameter of the circle?

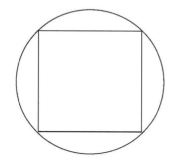

14. A circular garden with its center at point C contains an inner and outer portion, as shown in the figure. What is the area of the shaded part of the garden?

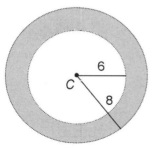

Plane Geometry

15. In the figure shown, *QRST* is a square with diagonals that meet at point *O*. The length of $\overline{OS}$ is $\sqrt{2}$ centimeters. What is the perimeter of *QRST*?

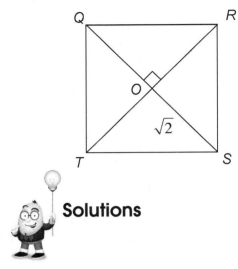

Solutions

1. The perimeter of *MNOPQ* is 84.

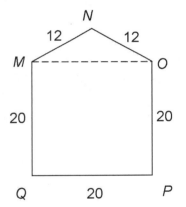

Determine the lengths of the missing sides. *MOPQ* is a square, so each side measures 20. The perimeter of the pentagon *MNOPQ* is 20 + 20 + 20 + 12 + 12, which equals 84.

2. The area of pentagon *MNOPQ* is $400 + 20\sqrt{11}$ square units.

To find the area of the figure, we first draw in the lengths of the missing sides.

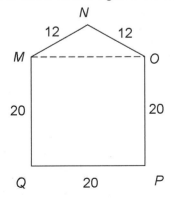

Chapter 9: Irregular and Multiple Figures

The figure can be divided into triangle *MNO* and square *MOPQ*. Find the area of these two shapes and add them together.

The area of *MOPQ* is 20 times 20, or 400. The area of $\triangle MNO$ is $\frac{1}{2}$ the base times the height of the triangle. To find the base and height, draw a perpendicular line from point *N* to the middle of $\overline{MO}$, as shown. We have labeled the perpendicular $\overline{NR}$, as shown. The perpendicular bisects $\overline{MO}$, so the base of the triangle *MNR* measures 10.

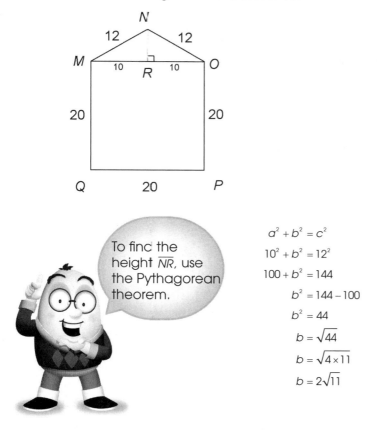

To find the height $\overline{NR}$, use the Pythagorean theorem.

$$a^2 + b^2 = c^2$$
$$10^2 + b^2 = 12^2$$
$$100 + b^2 = 144$$
$$b^2 = 144 - 100$$
$$b^2 = 44$$
$$b = \sqrt{44}$$
$$b = \sqrt{4 \times 11}$$
$$b = 2\sqrt{11}$$

The height of the triangle is $2\sqrt{11}$, and the base is 20. Use the formula for the area of a triangle:

$$A = \frac{1}{2}(b \times h)$$
$$= \frac{1}{2}\left(20 \times 2\sqrt{11}\right)$$
$$= \frac{1}{2}\left(40\sqrt{11}\right)$$
$$= 20\sqrt{11}$$

The area of the triangle is $20\sqrt{11}$, and the area of the square is 400. The area of pentagon *MNOPQ* is $400 + 20\sqrt{11}$ square units.

Plane Geometry

3. The area of the figure is 100 square units.

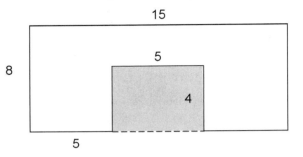

This figure contains a smaller rectangle inside a larger one. The smaller rectangle is shaded in the figure shown.

To find the area of the white part of the figure, subtract the area of the shaded rectangle from the area of the larger one. The area of the larger rectangle is 8 times 15, or 120 square units. The area of the shaded rectangle is 5 times 4, or 20 square units. The area of the larger rectangle minus the shaded portion is 120 – 20, or 100 units^2.

4. To walk from store *B* to store *E* using one of the routes shown, Bonnie must walk 39 blocks.

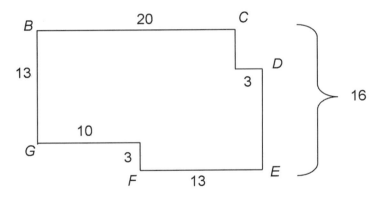

First, find the lengths of the missing sides. We know that the missing side between *G* and *F* measures 10 blocks, because there are 23 blocks in an east direction between building *B* and building *D*. Subtract the length of $\overline{FE}$ from 23 to find the missing length at the bottom of the figure: 23 – 13 = 10 blocks.

The distance from building *C* to building *E* is 16 blocks. We know this because there are 13 blocks between building *B* and building *G*. Building *F* is an additional 3 blocks farther south.

Bonnie could walk one of two routes. First, she could walk east from building *B* to building *C* and then south and east to building *D* and straight south to building *E*. This would be a total of 20 + 3 + 16, or 39 blocks. Alternatively, she could walk south 13 blocks from building *B* to building *G*, then 10 blocks east and another 3 blocks south to building *F*, followed by 13 blocks east from building *F* to building *E*. That route would be 13 + 10 + 3 + 13, or 39 blocks as well. Either way, Bonnie would walk 39 blocks.

Chapter 9: Irregular and Multiple Figures

5. The perimeter of $\triangle OHG$ is 32 meters.

Triangle *EGH* is a right triangle with two legs measuring 8 and 15 meters. Therefore, it is a special right triangle—one of the Pythagorean triples. Its sides measure 8, 15, and 17.

The length of hypotenuse $\overline{EG}$ is 17 meters. So, the length of each missing side of the triangle is half of 17, or 8.5 meters. The perimeter of $\triangle OHG$ is 8.5 + 8.5 + 15 = 32 meters.

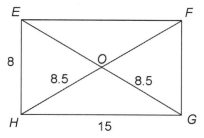

6. The measure of *s* is 69°.

The diagonals create two sets of vertical angles. The measure of $\angle QTR$ is 42°, and $\angle PTS$ is equal to $\angle QTR$. So, $\angle PTS$ also measures 42°.

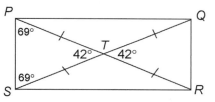

Using the rules of diagonals, we know that $\overline{PR}$ and $\overline{SQ}$ bisect each other. This tells us that $\overline{PT}$ is congruent to $\overline{ST}$. The triangle *PTS* is isosceles, and $\angle TPS$ is congruent to $\angle TSP$.

The triangle has two equal angles plus another that measures 42°. The three angles must add up to 180°. Use subtraction to find the value of *x*:

$$x + x + 42 = 180$$
$$2x + 42 = 180$$
$$2x = 180 - 42$$
$$2x = 138$$
$$x = 69$$

Plane Geometry

7. The value of *a* is 36°.

The figure shows that one angle measures 72°. The triangle is isosceles, so it has two equal angles. Since two angles measure 72°, subtract these from 180° to find the value of the third: 180 – 72 – 72 = 36.

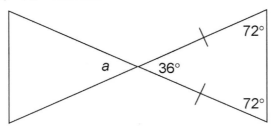

The 36° angle is a vertical angle with the angle labeled *a*. So, ∠*a* also measures 36°.

8. The length of $\overline{BD}$ is 12 inches.

The rectangle contains two right triangles. Both are 30-60-90 triangles. If ∠*BDC* measures 30°, then the measure of ∠*ADB* must be 60°. This means ∠*ABD* measures 30°.

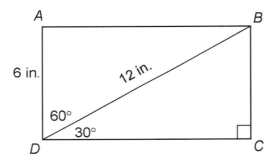

Once we know that this is a special right triangle, we can use its properties to determine the lengths of its sides. The sides of a 30-60-90 triangle are in a ratio of $1 : \sqrt{3} : 2$. The side across from the 30° angle is the smallest side; in this case, *AD* = 6 inches. Therefore, the hypotenuse $\overline{BD}$ must measure 2 times 6, or 12 inches.

9. The area of the swimming pool is 1,728 square feet.

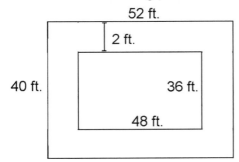

The outer dimensions of the walkway are 40 feet by 52 feet. The walkway measures 2 feet in width, so subtract to find the width of the pool: 40 feet – 2 feet – 2 feet = 36 feet in width, and 52 feet – 2 feet – 2 feet = 48 feet in length. The dimensions of the pool are 36 feet by 48 feet. Its area is 36 times 48, or 1,728 square feet.

10. The measure of $\overline{IJ}$ is 9.

We are shown one angle in the figure that measures 60°. Draw a line through the middle of the rhombus to create two triangles.

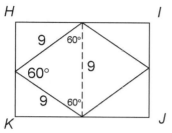

A rhombus has four congruent sides, so we know all sides measure 9. This means that the three angles of the triangle are equal. If the known angle measures 60°, and the remaining two angles are equal, they must measure 60° too.

The triangle is an equilateral triangle, with three equal angles and three equal sides. The side marked with the dotted line must measure 9. This means side $\overline{IJ}$ is also 9 units in length.

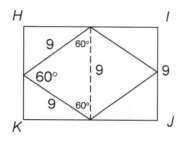

11. The area of circle R is 9π square feet.

Line segment $\overline{RO}$ bisects the base of the triangle, which measures 8 feet. This tells us that $ON = 4$ feet. We're also told that $RN = 5$ feet.

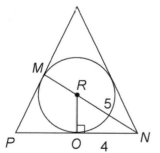

Notice that triangle RNO has two sides that measure 4 and 5 feet. The hypotenuse, $\overline{RN}$, is the side that measures 5. Therefore, this is a 3-4-5 Pythagorean triple. The remaining side must measure 3 feet:

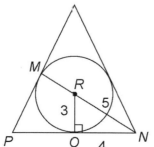

We now know the radius of circle R is 3 feet. Using this information, find the area of the circle:

$$A = \pi r^2$$
$$= \pi 3^2$$
$$= 9\pi$$

Plane Geometry

12. The measure of *z* is 45°.

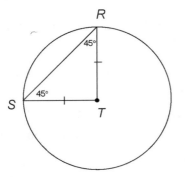

Circle *T* contains an inscribed triangle, *RTS*. The measure of ∠*T* is 90 degrees. The two legs of the triangle are congruent, because they are both radii of the circle. Therefore, m∠*S* ≅ m ∠*R*. Since the two angles add up to 90°, they each measure 45°.

13. The diameter of the circle measures $12\sqrt{2}$ units.

The square has sides that measure 12 units. Draw in the lengths of the sides and the diameter of the circle:

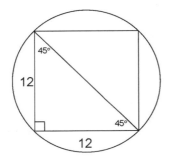

The diameter of the circle is also a diagonal of the inscribed square. According to the rules of diagonals, the diagonals of a square bisect the angles of the square. Therefore, the angles measure 45°, as shown.

We now have a 45-45-90 triangle, with sides in a ratio of $1:1:\sqrt{2}$. Since the legs of the triangle measure 12 units, the hypotenuse must measure $12\sqrt{2}$ units.

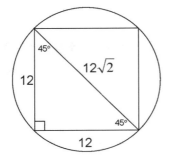

14. The shaded part of the garden has an area of 28π square units.

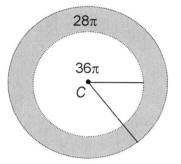

The area of the entire outer garden is π times 8^2, or 64π. The area of the inner garden is π times 6^2, or 36π. Subtract the area of the inner garden from the area of the entire outer garden to get the area of the shaded portion: 64π – 36π = 28π units^2.

15. The perimeter of *QRST* is 8 centimeters.

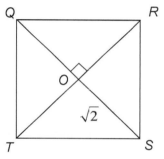

To find the perimeter of *QRST*, we must know the length of one side. This is possible if we recognize that the diagonals create special right triangles.

The rules of diagonals for squares state that the diagonals bisect the angles of the square. So, the diagonals create 8 angles each measuring 45 degrees:

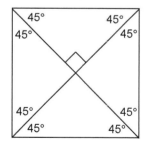

Therefore, we know that *RSO* is a 45-45-90 triangle. It has two congruent legs, as shown:

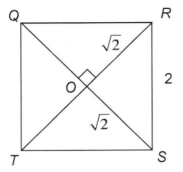

The sides of a 45-45-90 triangle are in the ratio $1:1:\sqrt{2}$. If the two legs measure $\sqrt{2}$, then the hypotenuse must measure $\sqrt{2} \times \sqrt{2}$.

Multiply the two square roots:

$$\sqrt{2} \times \sqrt{2} = \sqrt{4}$$
$$= 2$$

The length of the hypotenuse is 2 centimeters.

Add up the lengths of the sides to find the perimeter: $2 + 2 + 2 + 2 = 8$.

We've covered a lot of ground with 2-D shapes so far. Next, we'll move on to shapes in three dimensions.

Part 2

Solids

Chapter 10

Cubes

Hi! I'm egghead. I will teach the following concepts in this chapter:

Whct is a cube?
Faces, vertices, and edges
Dimensions
Volume
Units
Surface area

What is a cube?

Until now, all the shapes we've seen have been two-dimensional. These shapes only have two measurable dimensions: length and width. But there are also THREE-dimensional shapes, or 3-D shapes. Their dimensions are: length, width, and height.

Put on your 3-D glasses!

The 3-D figures in geometry are also called **solids.** One popular geometry solid is the cube.

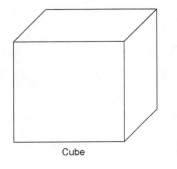

Cube

A **cube** is a solid figure with six equal, square faces.

Faces, vertices, and edges

Cubes in geometry generally don't have names. They do have some specific parts, however.

The square sides of a cube are called **faces.**

Every cube has six faces.

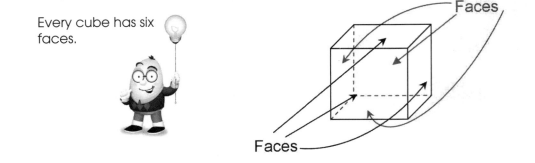

Faces

Faces

Chapter 10: Cubes

The points of a cube are called **vertices.** Just like the connection points of angles are called vertices, so are the connection points of cubes.

Every cube has eight vertices.

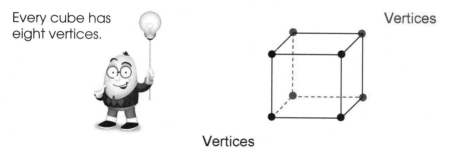

Vertices

Vertices

The sides of each square face are called **edges.**

Every cube has 12 edges.

Edges

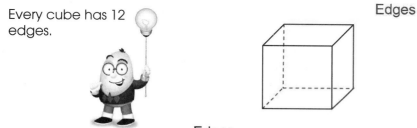

Edges

All of the edges of a cube are equal in length.

Dimensions

The most important dimension to know for a cube is the length of its edge.

Examples

The length of the edges of this cube is 4:

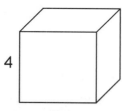

4

The edges of this cube measure 7:

7

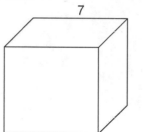

The edges of a cube are equal. You only need the measurment of one to know them all!

Volume

The **volume** of a cube is the amount of space that the cube takes up, or the amount that the cube can hold. The dotted lines show that this cube is about $\frac{1}{4}$ full with water.

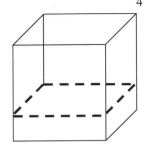

Formula

To find the volume of a cube, we multiply length times width times height.

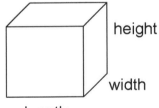

height

width

length

For a cube, those measures are always equal:

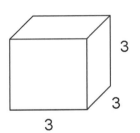

3

3

3

Remember, all edges are equal!

To find the volume, we multiply edge times edge times edge:

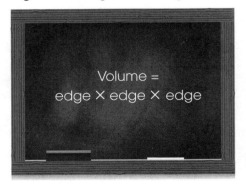

Volume =
edge × edge × edge

In shorthand, it is written this way:

You might also see it like this:

Again, that's just a short way of saying $e \times e \times e$.

Units

The volume of a shape is always given in cubic units. This is true not just for cubes, but for all 3-D shapes.

If the length of the edge of a cube is given in inches, the volume of the cube is in cubic inches. If the length of the edge is given in centimeters, the volume will be in cubic centimeters, and so on.

The volume can be abbreviated as units^3.

Examples

To find the volume of this cube, multiply edge × edge × edge:

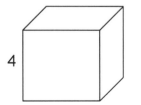

In this case, the edge is 4. The volume equals $4 \times 4 \times 4$, or 64 cubic units.

Solids

To find the volume of this cube, multiply edge × edge × edge:

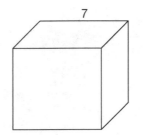

In this case, the edge is 7. The volume equals 7 × 7 × 7, or 343 cubic units.

Practice Questions

1. Find the volume of the cube shown.

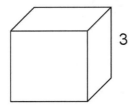

2. What is the volume of the cube shown?

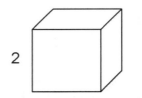

3. If the volume of the cube shown is 125 cubic centimeters, what is the length of edge *a*?

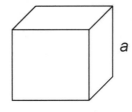

$V = 125 \text{ cm}^3$

4. What is the volume of the cube shown?

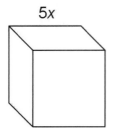

Chapter 10: Cubes

5. The edge of the cube shown measures 25 inches. Find its volume.

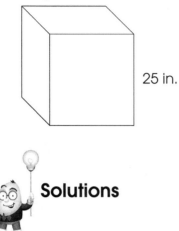

25 in.

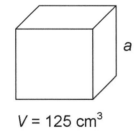 **Solutions**

1. To find the volume of a cube, multiply edge × edge × edge. The edge of this cube is 3, so the volume equals 3 × 3 × 3. The volume is 27 cubic units.

2. The edge of this cube is 2, so the volume equals 2 × 2 × 2. The volume is 8 cubic units.

3. The length of edge *a* is 5 centimeters.

a

$V = 125 \text{ cm}^3$

We know the volume is 125 cubic centimeters. Use the volume formula and work backwards to solve for *a*:

$$V = e^3$$
$$125 = a^3$$

In this case, to find *a*, we must take the cube root of 125. A cube root is similar to a square root, but it is the number that must be multiplied by itself three times to produce the result.

$$\sqrt[3]{125} = \sqrt[3]{a^3}$$
$$\sqrt[3]{5 \times 5 \times 5} = \sqrt[3]{a \times a \times a}$$
$$5 = a$$

To check the answer, multiply 5 times 5 times 5: the result is 125. Edge *a* measures 5 centimeters.

4. The volume of the cube is $125x^3$ cubic units.

5x

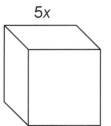

This was a tricky one. We actually don't know length of the edge. But we are given that the length is 5x. So, we plug 5x into our formula, and we multiply 5x times 5x times 5x, which equals $125x^3$. So, the volume of this cube is $125x^3$ cubic units.

5. The volume of the cube is 15,625 cubic inches.

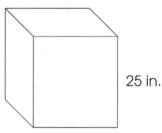

25 in.

We are given that the edge length measures 25 inches. Plug this into the volume formula:

$$V = e^3$$
$$= 25^3$$
$$= 25 \times 25 \times 25$$
$$= 15,625$$

The volume is 15,625 inches^3.

Surface area

Along with volume, another important measurement for cubes is **surface area.** Surface area is the area covered by the faces of a cube:

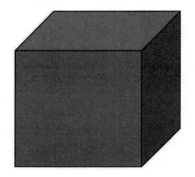

The shaded part shows the surface area.

Formula

Each cube has six faces that are squares. To find the surface area, we add up the areas of all six squares.

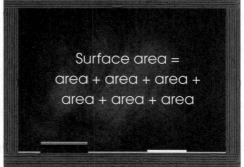

To find the area of one square, we multiply the edge by the edge:

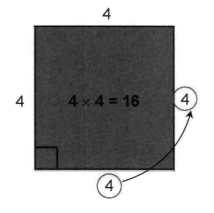

Since all squares on a cube are equal, we can multiply the area of one square by 6:

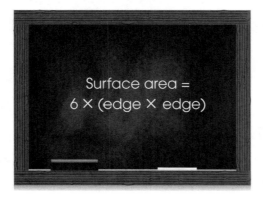

Solids

A short way to write that would be:

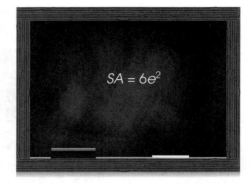

The $6e^2$ means $6 \times e \times e$. That's an easier way to write the formula. Surface area is always given in square units.

Examples

Here are some examples of how to find surface area.

To find the surface area of this cube, multiply $6 \times$ edge $\times$ edge:

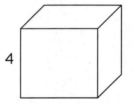

4

The edge of the cube measures 4. The area of one square face is 4×4, or 16. The surface area of the entire cube is 6×16, or 96 square units.

To find the surface area of this cube, multiply $6 \times$ edge $\times$ edge:

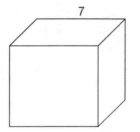

7

In this case, the edge is 7. The area of one square face is 7×7, or 49. The surface area of the entire cube is 6×49, or 294 square units.

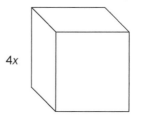

$4x$

For this example, the measurement of the edge is an unknown quantity. But we can still figure out the area. The length of the edge is given as $4x$. The area of one square face is $4x$ times $4x$, or $16x^2$. The surface area of the entire cube is 6 times $16x^2$, or $96x^2$.

egghead's Guide to Geometry

Practice Questions

1. What is the surface area of the cube shown?

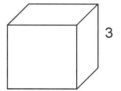

3

2. Find the surface area of the cube shown.

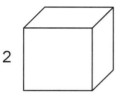

2

3. The area of one face of a cube is 144 mm^2, as shown in the figure. What is the surface area of the cube?

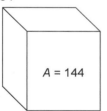

A = 144

4. If the surface area of the cube shown is 216 square feet, what is the length of its edge?

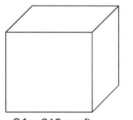

SA = 216 sq. ft.

5. The cube shown has edges of length 17x. Find its surface area.

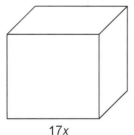

17x

Solutions

1. To find the surface area, multiply 6 × edge × edge. The edge of this cube measures 3, so the surface area equals 6 × 3 × 3. The surface area is 54 square units.

2. To find the surface area, multiply 6 × edge × edge. The edge of this cube measures 2, so the surface area equals 6 × 2 × 2. The surface area is 24 square units.

3. The surface area of the cube is 864 square millimeters.

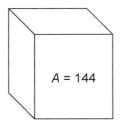

$A = 144$

We know that a cube is made up of six square faces. If the area of one of the faces is equal to 144 mm^2, we can multiply this by 6 to find the surface area of the cube: 144 × 6 = 864 mm^2.

4. The length of the edge is 6 feet.

$SA = 216$ sq. ft.

This is a tricky problem, but we can solve it by working backwards. We know the surface area of the cube is 216 square feet. So, that means $216 = 6e^2$. Divide both sides of the equation by 6, which gives us $36 = e^2$. Take the square root of 36 to find the value of e. The edge measures 6 feet.

$$SA = 6e^2$$
$$216 = 6e^2$$
$$\frac{216}{6} = \frac{6e^2}{6}$$
$$36 = e^2$$
$$e = 6$$

5. The surface area of the cube is $1,734x^2$ square units.

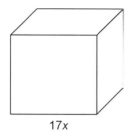

17x

Use the surface area formula:

$$SA = 6e^2$$
$$= 6 \times (17x)^2$$
$$= 6 \times (17 \times 17 \times x \times x)$$
$$= 6 \times (289 \times x^2)$$
$$= 1,734x^2$$

Chapter Review

1. What is the volume, in cubic units, of the cube shown?

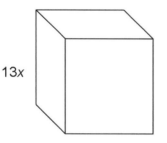

13x

2. What is the surface area, in square units, of the cube shown?

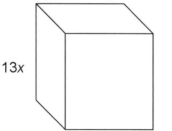

13x

3. What is the volume of the cube shown?

5

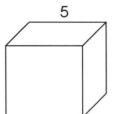

4. What is the surface area of the cube shown?

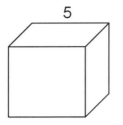

5. What are the volume and surface area of the cube shown?

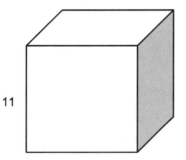

6. What is the volume of the cube shown?

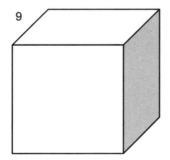

7. What is the surface area of the cube shown?

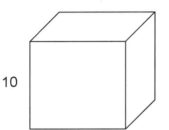

egghead's Guide to Geometry

8. The cube shown has a side length of 15 inches. Find its surface area.

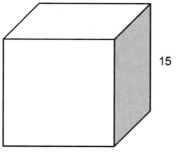

15

9. What is the volume of the cube shown, with side length 14 millimeters?

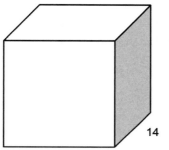

14

10. The surface area of a cube is 1,944 square inches, as shown. Find the length of its sides.

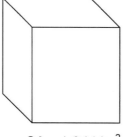

$SA = 1{,}944$ in.^{2}

11. A shed in the shape of a cube has dimensions of length 22 feet by 22 feet by 22 feet. A wall is built in the middle of the shed, as shown by the dotted line. The wall divides the shed into two rooms of equal size. What is the volume of one of the rooms?

22 feet

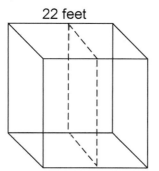

Solids

12. Amber has a cube-shaped container of side length 49 millimeters, as shown. She fills the cube with liquid, and then pours the contents of the cube into 10 identical test tubes. Each test tube contains the same amount of liquid. What volume of liquid does each test tube hold?

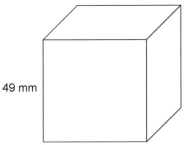

49 mm

13. Three storage buildings in the shape of a cube each have dimensions 19 meters by 19 meters by 19 meters, as shown. Each building is painted on 5 of its 6 sides. What is the total painted surface area of the three buildings?

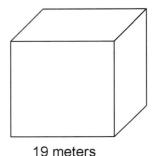

19 meters

14. A water tank in the shape of a cube has sides that measure 35 inches. The tank is filled exactly one-third full, as shown by the dotted line. What is the volume of water in the tank?

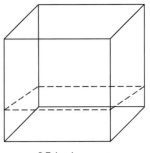

35 inches

15. A cube with sides measuring 20 centimeters long is divided into 8 smaller cubes of equal size, as shown in the figure. What is the volume of one of the smaller cubes?

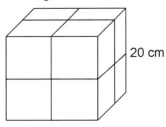

20 cm

Chapter 10: Cubes

Solutions

1. The volume of the cube is $2,197x^3$ cubic units.

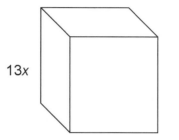

13x

Using the volume formula, multiply edge × edge × edge. In this case, we multiply $13x$ times $13x$ times $13x$. The volume is $2,197x^3$ cubic units.

2. The surface area of the cube is $1,014x^2$ square units.

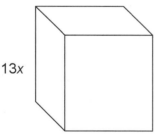

13x

First, find the area of one face of the cube: $13x$ times $13x$ equals $169x^2$. Multiply this figure by 6:

$$SA = 6e^2$$
$$= 6 \times (13x)^2$$
$$= 6 \times 169x^2$$
$$= 1,014x^2$$

3. To find the volume of a cube, multiply edge × edge × edge. The edge of this cube equals 5, so the volume equals 5 × 5 × 5. The volume is 125 cubic units.

4. To find the surface area, we multiply 6 × edge × edge. The edge of this cube equals 5, so the surface area equals 6 × 5 × 5. The surface area is 150 square units.

5. To find the volume, multiply edge × edge × edge. The volume measures 10 × 10 × 10, or 1,000 cubic units. To find the surface area, multiply 6 × edge × edge. The surface area is 6 × 10 × 10, or 600 square units.

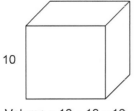

10

Volume = 10 × 10 × 10

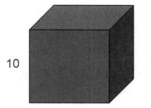

10

Surface Area = 6 × 10 × 10

Solids

6. The edge of this cube measures 11 units, so the volume equals 11 × 11 × 11.

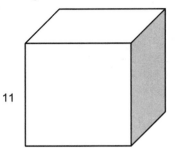

The volume is 1,331 cubic units.

7. The surface area is 486 units^2.

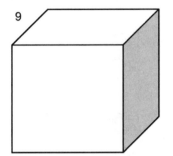

To find the surface area, we multiply 6 × edge × edge. The edge of this cube equals 9, so the surface area equals 6 × 9 × 9, or 486 square units.

8. The surface area is 1,350 square inches.

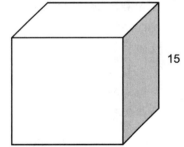

The edge of this cube measures 15 inches. The surface area therefore equals 6 × 15 × 15, or 1,350 square inches.

9. The volume is 2,744 mm^3.

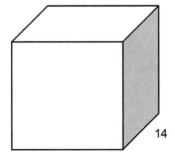

The edge of this cube measures 14 millimeters, so its volume equals 14 × 14 × 14, or 2,744 cubic millimeters.

egghead's Guide to Geometry

10. The sides of the cube measure 18 inches.

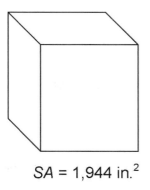

Using the surface area formula, work backward to find the length of the side:

$$SA = 6e^2$$
$$1{,}944 = 6e^2$$
$$\frac{1{,}944}{6} = e^2$$
$$324 = e^2$$
$$\sqrt{324} = \sqrt{e^2}$$
$$18 = e$$

The sides measure 18 inches.

11. The volume of one room is 5,324 cubic feet.

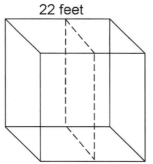

22 feet

First, find the volume of the entire shed. The volume is 22 times 22 times 22, or 10,648 cubic feet. Divide this number in half to find the volume of one room: 10,648 divided by 2 equals 5,324 cubic feet.

12. The volume of liquid in each test tube is 11,764.9 cubic millimeters.

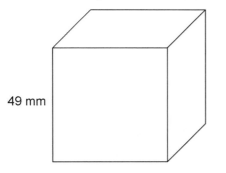

49 mm

Each test tube contains the same amount of liquid. So, we would find the total volume of liquid and divide this by 10 to find the volume in each test tube.

The total volume of liquid is 49 × 49 × 49, or 117,649 cubic millimeters. Divide this number by 10: 117,649 × 10 = 11,764.9 mm^3.

Solids

13. The painted surface area of the three buildings is 5,415 square meters.

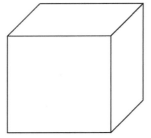

19 meters

First, find the surface area of one face of one building. The surface area of one face is 19 times 19, or 361 square meters.

Then, find the painted surface area of one building. Five of the sides of the building are painted, so the painted surface area is 361 × 5, or 1,805 square meters.

Multiplying this times 3, we see that the painted surface area of all three buildings is 5,415 square meters.

14. The volume of water in the tank is approximately 14,291.67 cubic inches.

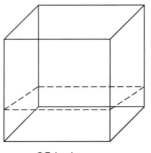

35 inches

The total volume of the tank is 35 × 35 × 35, or 42,875 cubic inches. To find the volume of water, divide 42,875 by 3. The tank is one-third full, so it contains about 14,291.67 cubic inches of water.

15. The volume of each small cube is 1,000 cubic centimeters.

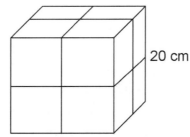

20 cm

This cube has sides of length 20 centimeters. So, its total volume is 20 × 20 × 20, or 8,000 cm³. There are 8 small cubes in the figure. The volume of one small cube is 8,000 divided by 8, or 1,000 cm³.

Chapter 11

Rectangular Solids

Hi! I'm egghead. I will teach the following concepts in this chapter:

What is a rectangular solid?
Properties of rectangular
 solids
Dimensions
Volume
Surface area

What is a rectangular solid?

Rectangular solids are similar to cubes. A **rectangular solid** is a three-dimensional (3-D) figure with six rectangular faces. You've seen them before. They include shoe boxes, cereal boxes, and even bricks.

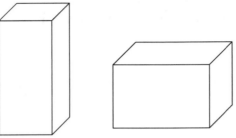

Properties of rectangular solids

Like cubes, rectangular solids in geometry don't usually have names. But they do have the same specific parts as cubes do.

Every rectangular solid has six surfaces, or faces.

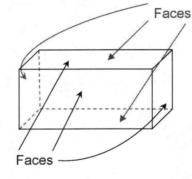

Faces

Faces

Every rectangular solid has eight points, or vertices.

Vertices

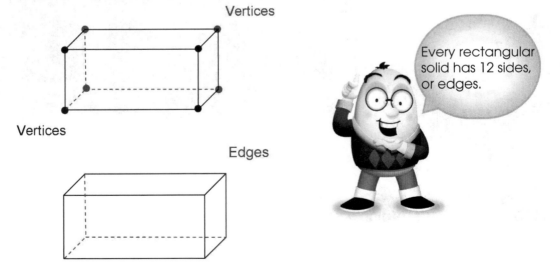

Vertices

Edges

Every rectangular solid has 12 sides, or edges.

Edges

egghead's Guide to Geometry

Dimensions

Like other 3-D shapes, rectangular solids have three dimensions: length, width, and height.

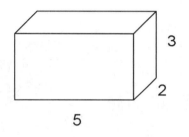

Examples

This rectangular solid has a length of 5 inches, a width of 2 inches, and a height of 3 inches:

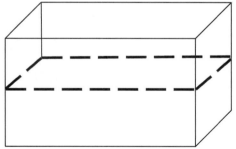

Volume

These dimensions come in handy for measuring volume. The volume of a rectangular solid is the amount of space that the rectangular solid takes up, or the amount that the rectangular solid can hold.

Formula

To find the volume of a rectangular solid, we multiply length times width times height.

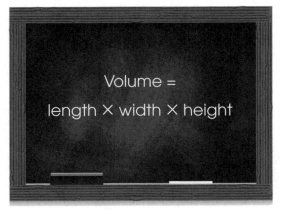

Volume =
length × width × height

In shorthand, it's written like this:

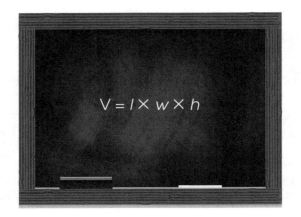

V = l × w × h

Examples

To find the volume of this rectangular solid, multiply length × width × height:

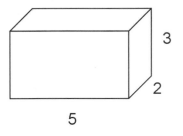

The length of this rectangular solid is 5. The width is 2, and the height is 3. The volume measures 5 × 2 × 3, or 30 cubic units.

To find the volume of this rectangular solid, multiply length × width × height:

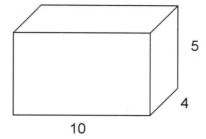

In this case, the length is 10, the width is 4, and the height is 5. The volume measures 10 × 4 × 5, or 200 cubic units.

Practice Questions

1. What is the volume of the rectangular solid shown?

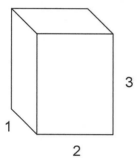

2. What is the volume of the rectangular solid shown?

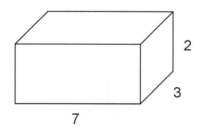

3. You're moving, and the moving company sends you 4 boxes that are each 7 inches long, 12 inches wide, and 6 inches tall. What is the volume of all four boxes?

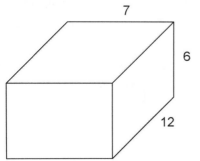

Solids

4. Sharon has 3 wooden boxes that are each 20 centimeters long, 9 centimeters wide, and 11 centimeters high. What is the total volume of the 3 boxes?

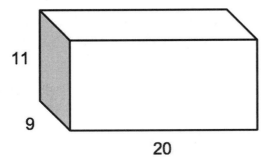

5. Stan buys 5 glass containers. Each container has the dimensions shown. If the dimensions are given in centimeters, what is the volume, in cubic centimeters, of all 5 containers?

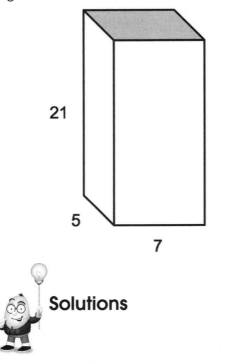

Solutions

1. To find the volume of a rectangular solid, multiply length × width × height. The length of this rectangular solid is 2. The width is 1, and the height is 3. The volume measures 2 × 1 × 3, or 6 cubic units.

2. To find the volume of a rectangular solid, multiply length × width × height. The length of this rectangular solid is 7. The width is 3, and the height is 2. The volume measures 7 × 3 × 2, or 42 cubic units.

Chapter 11: Rectangular Solids

3. The combined volume of all 4 boxes is 2,016 cubic inches.

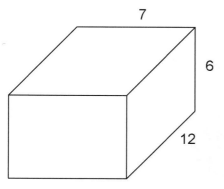

Multiply the length, width, and height of one box: 7 × 12 × 6 = 504 inches^3. Then, multiply that number by 4 to determine the volume of all 4 boxes: 504 times 4 equals 2,016 cubic inches.

4. The total volume of the 3 boxes is 5,940 centimeters^3.

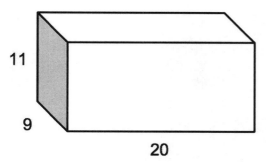

The length of one box is 20 centimeters. The width is 9 centimeters, and the height is 11 centimeters. The volume of one box equals 20 × 9 × 11, or 1,980 cm^3. Multiply this by 3 to find the volume of all 3 boxes: 1,980 times 3 equals 5,940 cubic centimeters.

5. The volume of all 5 containers is 3,675 cubic centimeters.

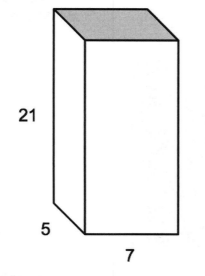

The length of each container is 7. The width is 5, and the height is 21. The volume of one container is 7 × 5 × 21, or 735 cm^3. The combined volume of 5 containers is 735 times 5, or 3,675 cubic centimeters.

Surface area

With rectangular solids, we also measure surface area.

Surface area is the area covered by the faces of a rectangular solid:

Formula

To find the surface area, we add up the area of each face of the figure. The formula can be written this way:

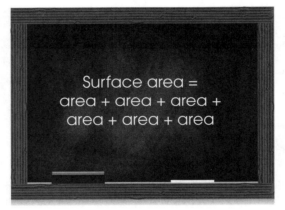

Surface area =
area + area + area +
area + area + area

To find the area of each rectangular face, we multiply the sides of each rectangle. First we find all six areas, and then we add them. We add together length × width, plus length × width again, plus length × height, plus length × height again, plus width × height, plus width × height again.

Here's another way to write the formula to help you remember:

Surface area =
2 × (length × width) +
2 × (length × height) +
2 × (width × height)

Chapter 11: Rectangular Solids

You might also see it written this way:

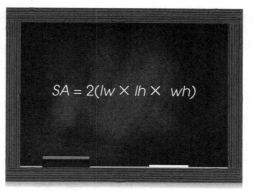

$$SA = 2(lw \times lh \times wh)$$

Finding the surface area of rectangular solids can be challenging, but fortunately, surface area and volume are the only formulas you really need to know. On standardized tests, most questions about rectangular solids ask about surface area or volume—that's it!

Examples

Here are some examples of how to find surface area.

To find the surface area of this rectangular solid, calculate $2lw + 2lh + 2wh$:

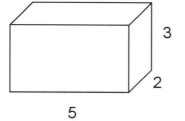

The length of this rectangular solid is 5. The width is 2, and the height is 3. Substitute these into the surface area formula:

$$
\begin{aligned}
SA &= 2lw + 2lh + 2wh \\
&= 2(5 \times 2) + 2(5 \times 3) + 2(2 \times 3) \\
&= 2(10) + 2(15) + 2(6) \\
&= 20 + 30 + 12 \\
&= 62
\end{aligned}
$$

To find the surface area of this rectangular solid, calculate $2lw + 2lh + 2wh$:

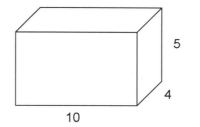

In this case, the length of this rectangular solid is 10. The width is 4, and the height is 5. Substitute these into the surface area formula:

$SA = 2lw + 2lh + 2wh$
$= 2(10 \times 4) + 2(10 \times 5) + 2(4 \times 5)$
$= 2(40) + 2(50) + 2(20)$
$= 80 + 100 + 40$
$= 220$

The surface area is 220 square units.

Practice Questions

1. What is the surface area of the rectangular solid shown?

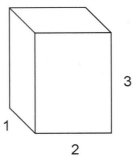

2. Find the surface area of the rectangular solid shown.

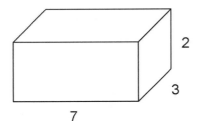

3. The box shown is 7 inches long, 12 inches wide, and 6 inches tall. What is its surface area?

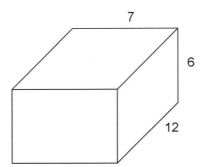

4. A rectangular solid measures 14 centimeters by 7 centimeters by 8 centimeters, as shown. Find its surface area.

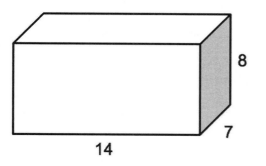

5. The rectangular solid shown measures 6 feet by 4 feet by 12 feet. What is the surface area of the solid?

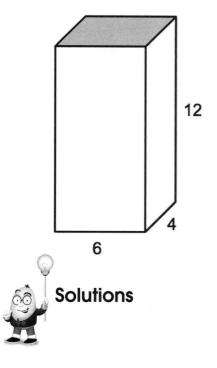

Solutions

1. The length of this rectangular solid is 2. The width is 1, and the height is 3. Substitute these into the surface area formula.

$$SA = 2lw + 2lh + 2wh$$
$$= 2(2 \times 1) + 2(2 \times 3) + 2(1 \times 3)$$
$$= 2(2) + 2(6) + 2(3)$$
$$= 4 + 12 + 6$$
$$= 22$$

Solids

2. To find the surface area of this rectangular solid, calculate $2lw + 2lh + 2wh$. The length of this rectangular solid is 7. The width is 3, and the height is 2. Substitute these into the surface area formula.

The surface area is $2(7 \times 3) + 2(7 \times 2) + 2(3 \times 2)$. That equals $42 + 28 + 12$. The surface area equals 82 square units.

3. The surface area of the box is 396 square inches.

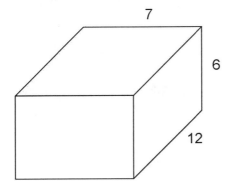

Use the surface area formula, $2lw + 2lh + 2wh$. In this case, we substitute 7 inches for length, 12 inches for width, and 6 inches for height:

$$
\begin{aligned}
SA &= 2lw + 2lh + 2wh \\
&= 2(7 \times 12) + 2(7 \times 6) + 2(12 \times 6) \\
&= 2(84) + 2(42) + 2(72) \\
&= 168 + 84 + 144 \\
&= 396
\end{aligned}
$$

4. The surface area of the box is 532 cm^2.

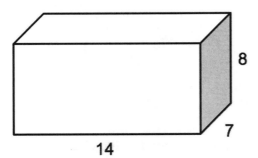

The length of this rectangular solid is 14 centimeters. The width is 7 centimeters, and the height is 8 centimeters. The surface area is $2(14 \times 7) + 2(14 \times 8) + 2(7 \times 8)$. That equals $196 + 224 + 112$, or 532 square centimeters.

5. The surface area of the solid is 288 square feet.

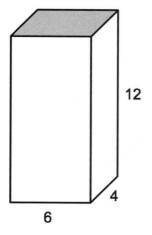

The length of this rectangular solid is 6 feet. The width is 4 feet, and the height is 12 feet. Substitute these into the surface area formula, $2lw + 2lh + 2wh$. The surface area is $2(6 \times 4) + 2(6 \times 12) + 2(4 \times 12)$. That equals $48 + 144 + 96$, or 288 square feet.

Chapter Review

1. What is the volume of the rectangular solid shown?

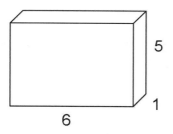

2. What is the surface area of the rectangular solid shown?

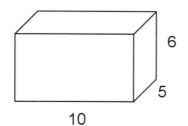

Solids

3. What is the volume of the rectangular solid shown?

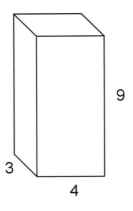

4. What is the surface area of the rectangular solid shown?

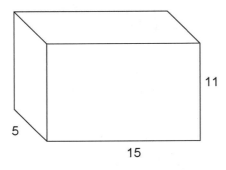

5. Find the volume and surface area of the rectangular solid shown.

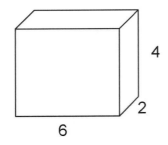

6. What is the volume of the rectangular solid shown?

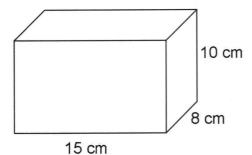

egghead's Guide to Geometry

Chapter 11: Rectangular Solids

7. The figure shown is a rectangular solid measuring 18 inches by 9 inches by 27 inches, as shown. Find its surface area.

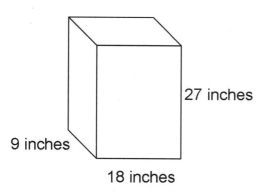

27 inches

9 inches

18 inches

8. Josie cuts a block of wood to the dimensions shown in the figure. If the dimensions are given in centimeters, what is the surface area, in square centimeters, of the figure?

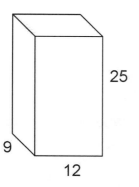

25

9

12

9. A metal container in the shape of a rectangular solid is 24 inches long, 8 inches wide, and 16 inches high. The container is used to package sand for shipping. What is the maximum volume, in cubic inches, of sand that the container can hold?

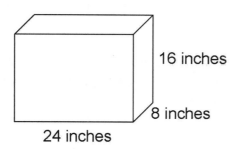

16 inches

8 inches

24 inches

Solids

10. A small glass tank is filled exactly half full with water, as shown by the dotted lines in the figure. What is the volume, in cubic centimeters, of water in the tank?

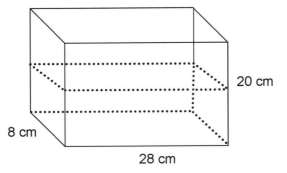

20 cm

8 cm

28 cm

11. A rectangular solid is shaded on exactly three sides, as shown in the figure. The remaining three sides, not seen in the figure, are not shaded. What is the total surface area of the shaded sides of the figure?

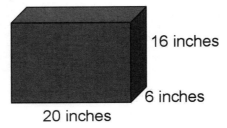

16 inches

6 inches

20 inches

12. Hazel has 6 gift boxes in the dimensions shown in the figure. What is the total volume, in cubic centimeters, of all 6 gift boxes?

12 cm

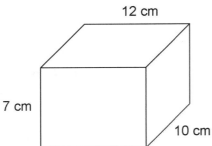

7 cm

10 cm

13. A dunking tank for a carnival is built with the dimensions shown. The tank is filled with water to a level of 4 feet, as shown by the dotted line. The height of the tank is 10 feet. How much more water, in cubic feet, would it take to fill the entire tank?

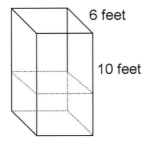

6 feet

10 feet

8 feet

egghead's Guide to Geometry

14. Meredith buys a trunk to place at the foot of her bed. The interior of the trunk measures 30 inches by 8 inches by 22 inches, as shown. In the trunk, Meredith stores ceramic figures packaged individually in boxes that each take up 220 square inches of room. How many ceramic figures can Meredith fit in the trunk?

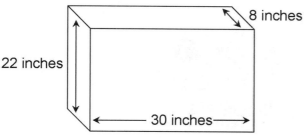

22 inches

8 inches

30 inches

15. A solid block of putty is formed into a rectangular solid with measurements 100 millimeters by 40 millimeters by 60 millimeters. Lamar cuts the putty in half along its length and height to form 4 equal blocks of putty, as shown. What is the surface area of one of the small blocks?

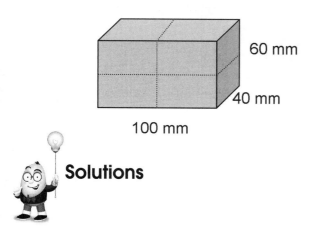

60 mm

40 mm

100 mm

Solutions

1. To find the volume of a rectangular solid, multiply length × width × height. The length of this rectangular solid is 6. The width is 1, and the height is 5. The volume measures 6 × 1 × 5, or 30 cubic units.

2. To find the surface area of this rectangular solid, calculate $2lw + 2lh + 2wh$. The length of this rectangular solid is 10. The width is 5, and the height is 6. Substitute these into the surface area formula.

The surface area is 2(10 × 5) + 2(10 × 6) + 2(5 × 6). That equals 100 + 120 + 60. The surface area is 280 square units.

3. To find the volume of a rectangular solid, multiply length × width × height. The length of this rectangular solid is 4. The width is 3, and the height is 9. The volume measures 4 × 3 × 9, or 108 cubic units.

Solids

4. The length of this rectangular solid is 15. The width is 5, and the height is 11. Substitute these into the surface area formula.

The surface area is 2(15 × 5) + 2(15 × 11) + 2(5 × 11). That equals 150 + 330 + 110. The surface area is 590 square units.

5.

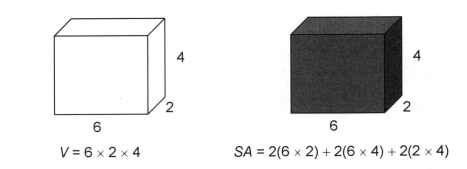

$$V = 6 \times 2 \times 4 \qquad\qquad SA = 2(6 \times 2) + 2(6 \times 4) + 2(2 \times 4)$$

To find the volume, multiply length × width × height. The volume measures 6 × 2 × 4, or 48 cubic units. To find the surface area, calculate 2*lw* + 2*lh* + 2*wh*. The surface area is 2(6 × 2) + 2(6 × 4) + 2(2 × 4), or 24 + 48 + 16. The surface area is 88 square units.

6. The volume of the rectangular solid is 1,200 cubic centimeters.

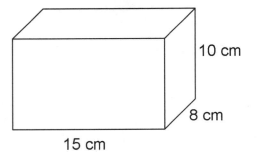

The length of this rectangular solid is 15 centimeters. The width is 8 centimeters, and the height is 10 centimeters. The volume measures 15 × 8 × 10, or 1,200 cubic centimeters.

7. The surface area of the figure is 1,782 square inches.

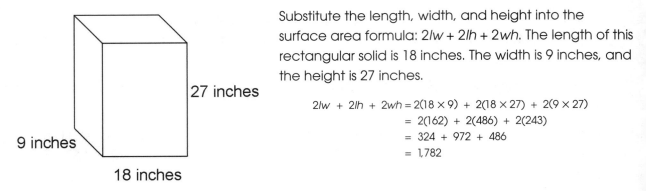

Substitute the length, width, and height into the surface area formula: 2*lw* + 2*lh* + 2*wh*. The length of this rectangular solid is 18 inches. The width is 9 inches, and the height is 27 inches.

$$\begin{aligned}
2lw + 2lh + 2wh &= 2(18 \times 9) + 2(18 \times 27) + 2(9 \times 27) \\
&= 2(162) + 2(486) + 2(243) \\
&= 324 + 972 + 486 \\
&= 1{,}782
\end{aligned}$$

egghead's Guide to Geometry

8. The surface area of the figure is 1,266 cm^2.

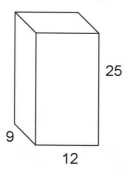

The block of wood measures 12 centimeters by 9 centimeters by 25 centimeters. To find the surface area, use the formula:

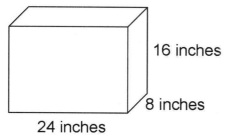

$$2lw + 2lh + 2wh = 2(12 \times 9) + 2(12 \times 25) + 2(9 \times 25)$$
$$= 2(108) + 2(300) + 2(225)$$
$$= 216 + 600 + 450$$
$$= 1,266$$

9. The container can hold a maximum of 3,072 cubic inches of sand.

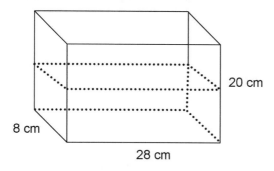

The container measures 24 by 8 by 16 inches. Multiply to find the volume: 24 inches times 8 inches times 16 inches is 3,072 in^3.

10. The volume of water in the tank is 2,240 cubic centimeters.

We are told that the tank is exactly half full with water. The volume of the entire tank is 28 × 8 × 20, or 4,480 cubic centimeters. Half of this amount is 2,240 cm^3.

Solids

11. The total surface area of the shaded sides is 536 in^2.

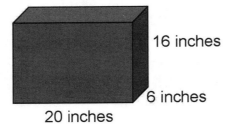

16 inches

6 inches

20 inches

The question tells us that the figure is shaded only on three sides. The remaining sides are not shaded. To find the surface area of the shaded sides, add up the area of each one.

The first shaded side is 20 by 16 inches, or 320 square inches. The second shaded side is 6 by 16, or 96 square inches. The third side, at the top of the figure, has an area of 20 by 6, or 120 square inches.

The total surface area of these sides is 320 + 96 + 120, or 536 square inches.

12. The total volume of all 6 gift boxes is 5,040 cubic centimeters.

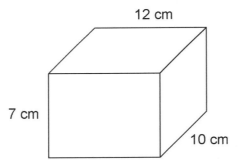

12 cm

7 cm

10 cm

Each box measures 12 centimeters by 10 centimeters by 7 centimeters. The volume of each box is therefore 12 times 10 times 7, or 840 cubic centimeters. The volume of 6 boxes is 5,040 cm^3.

13. To fill the entire tank, it would take an additional 288 cubic feet of water.

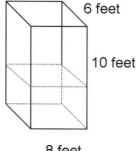

6 feet

10 feet

8 feet

The tank is already filled to a level of 4 feet, as shown by the dotted line. So, the part of the tank remaining to be filled measures 8 feet by 6 feet by 6 feet. The volume of water it would take to fill the tank is 8 × 6 × 6, or 288 ft^3.

You could also solve this problem by subtracting the volume of water in the tank from the volume of the whole tank. The volume of the entire tank is 8 feet times 6 feet times 10 feet, or 480 cubic feet. The volume of water in the tank is 8 feet by 6 feet by 4 feet, or 192 cubic feet. The volume remaining is 480 – 192, or 288 ft^3.

14. Meredith can fit 24 ceramic figures in the trunk.

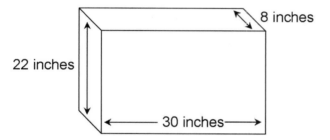

8 inches

22 inches

30 inches

First, find the volume of the trunk. The interior of the trunk measures 30 inches by 8 inches by 22 inches, so its volume is 30 times 8 times 22, or 5,280 square inches. Each ceramic figure takes up 220 square inches of room. Divide the 5,280 by 220 to determine the number of figures that can fit in the trunk: 5,280 divided by 220 is 24.

Solids

15. The surface area of one of the small blocks is 9,400 mm^2.

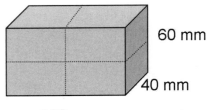

60 mm

40 mm

100 mm

Lamar makes one cut halfway along the length of the putty and another cut halfway along its height. As the figure shows, the width of the small blocks remains the same as that of the large block. Each small block therefore measures 50 mm by 40 mm by 30 mm.

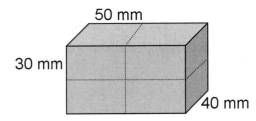

50 mm

30 mm

40 mm

To find the surface area, substitute these values into the formula:

$$
\begin{aligned}
2lw + 2lh + 2wh &= 2(50 \times 40) + 2(50 \times 30) + 2(40 \times 30) \\
&= 2(2,000) + 2(1,500) + 2(1,200) \\
&= 4,000 + 3,000 + 2,400 \\
&= 9,400
\end{aligned}
$$

egghead's Guide to Geometry

Chapter 12

Cylinders

Hi! I'm egghead. I will teach the following concepts in this chapter:

What is a cylinder?
Dimensions
Volume
Measuring surface area

What is a cylinder?

Along with cubes and rectangular solids, another three-dimensional shape is the **cylinder**.

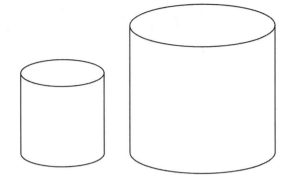

A cylinder is a solid figure made up of two equal bases joined by a curved side. The bases are usually circles.

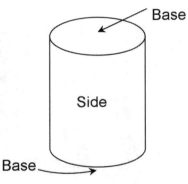

Dimensions

First, circular cylinders have a radius. The radius of a cylinder is the distance from the center of the figure to its edge. It is usually shown drawn on the base of the cylinder.

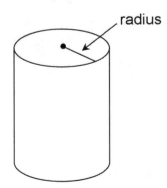

Chapter 12: Cylinders

Second, cylinders also have a height measure:

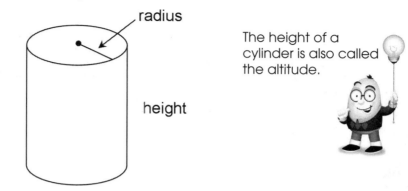

radius

height

The height of a cylinder is also called the altitude.

Examples

This is a cylinder with radius 3 and height 9:

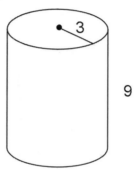

3

9

This is a cylinder with radius 2 and height 5:

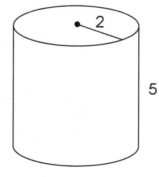

2

5

Volume

We use the radius and height of a cylinder to find its volume.

The **volume** is the amount of space that the cylinder takes up, or the amount that it can hold.

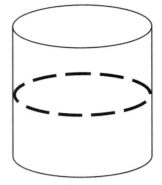

To find the volume of a cylinder, multiply the area of the base times the height.

The shaded section shows the area of the base.

radius

height

Remember how to find the area of a circle? We multiply that by the height of the cylinder.

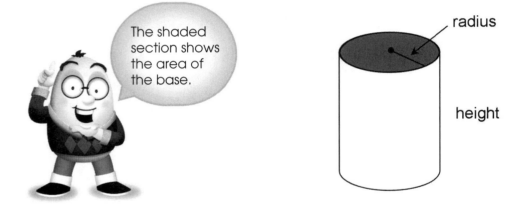

$$A = \pi \times r \times r$$

Chapter 12: Cylinders

In shorthand, it's written like this:

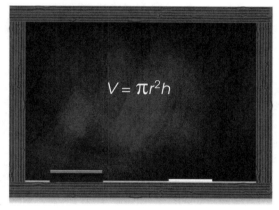

Examples

To find the volume of this cylinder, we multiply the area of the base times the height.

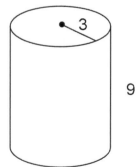

The cylinder has a radius of 3 and a height of 9.
Volume = $\pi r^2 h$:

$$\text{Volume} = \text{area of base} \times \text{height}$$
$$= \pi \times r \times r \times h$$
$$= \pi \times 3 \times 3 \times 9$$
$$= 81\pi$$

The volume of the cylinder is 81π cubic units.

Solids

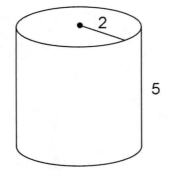

This cylinder has a radius of 2 and a height of 5. Multiply $\pi \times r \times r \times h$:

$$\text{Volume} = \text{area of base} \times \text{height}$$
$$= \pi \times r \times r \times h$$
$$= \pi \times 2 \times 2 \times 5$$
$$= 20\pi$$

The volume of this cylinder is 20π cubic units.

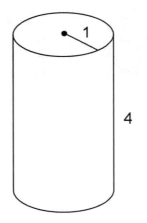

This cylinder has a radius of 1 and a height of 4.

$$\text{Volume} = \text{area of base} \times \text{height}$$
$$= \pi \times r \times r \times h$$
$$= \pi \times 1 \times 1 \times 4$$
$$= 4\pi$$

The volume of the cylinder is 4π cubic units.

Practice Questions

1. Find the volume of the cylinder shown.

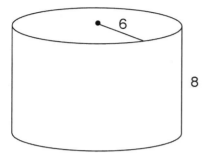

2. Find the volume of the cylinder shown.

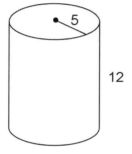

3. The cylinder shown has a diameter of 10 inches and a height of 14 inches. What is its volume?

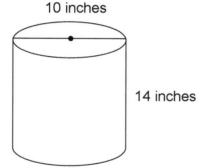

4. If a cylinder has a radius of 12 centimeters and a height of 14 centimeters, what is its volume?

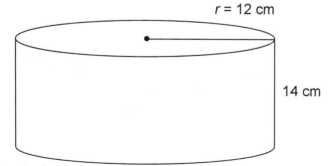

Solids

5. If a cylinder has a diameter of 8 centimeters and a height of 20 centimeters, what is its volume?

8 cm

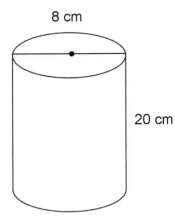

20 cm

 **Solutions**

1. To find the volume, we multiply $\pi \times r \times r \times h$:

$$\text{Volume} = \text{area of base} \times \text{height}$$
$$= \pi \times r \times r \times h$$
$$= \pi \times 6 \times 6 \times 8$$
$$= 288\pi$$

The volume is 288π cubic units.

2. The volume is 300π cubic units.

$$\text{Volume} = \text{area of base} \times \text{height}$$
$$= \pi \times r \times r \times h$$
$$= \pi \times 5 \times 5 \times 12$$
$$= 300\pi$$

3. The volume of the cylinder is 350π cubic inches.

10 inches

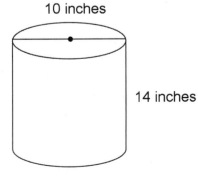

14 inches

The diameter of the cylinder is 10 inches, so its radius is 5 inches. Substitute 5 for *r* and 14 for *h* into the volume formula:

$$\text{Volume} = \text{area of base} \times \text{height}$$
$$= \pi \times r \times r \times h$$
$$= \pi \times 5 \times 5 \times 14$$
$$= 350\pi$$

egghead's Guide to Geometry

4. The volume of the cylinder is 2,016π cubic centimeters.

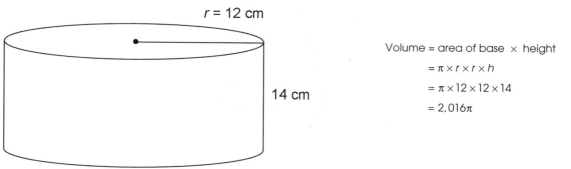

r = 12 cm

14 cm

Volume = area of base × height
= π × r × r × h
= π × 12 × 12 × 14
= 2,016π

5. The volume of the cylinder is 320π cubic centimeters.

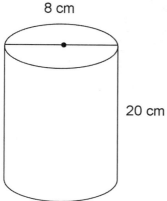

8 cm

20 cm

Again, we can find the radius by dividing the diameter by 2. The diameter of the cylinder is 8 centimeters, so its radius is 4 centimeters.

Volume = area of base × height
= π × r × r × h
= π × 4 × 4 × 20
= 320π

Measuring surface area

As with cubes and rectangular solids, we can also measure the surface area of cylinders.

The **surface area** is the area covered by the surfaces of the cylinder.

These are the two bases and the curved side.

Solids

To measure the surface area, we first take the area of the base:

$$A = \pi \times r \times r$$

Then we multiply that by two.

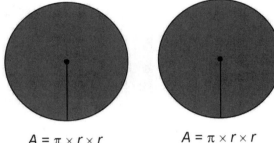

$$A = \pi \times r \times r \qquad A = \pi \times r \times r$$

Then we add the area of the curved side:

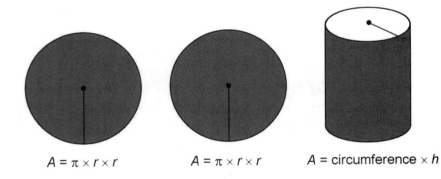

$$A = \pi \times r \times r \qquad A = \pi \times r \times r \qquad A = \text{circumference} \times h$$

To find the area of the curved side, we first need to know its circumference. To find the circumference, we use the formula $C = 2\pi r$. Since a cylinder has two circular bases, we can use the radius of the base to calculate the circumference of the curved side.

Then we multiply the circumference times the height of the cylinder to get the area of the curved side.

egghead's Guide to Geometry

Chapter 12: Cylinders

Here is what it looks like very simply:

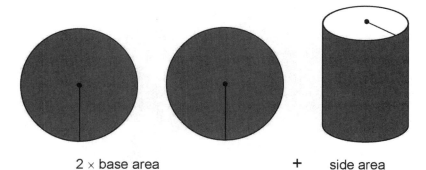

2 × base area   +   side area

This is how you would write the formula:

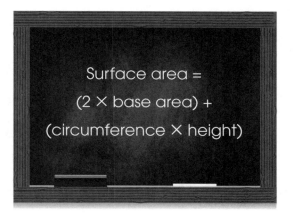

Surface area =
(2 × base area) +
(circumference × height)

Let's shorten that up a bit:

$SA = (2 \times \pi \times r \times r) + (2 \times \pi \times r \times h)$

Hey, that's pretty easy to remember!

Solids

It can also be written super fancy, like this:

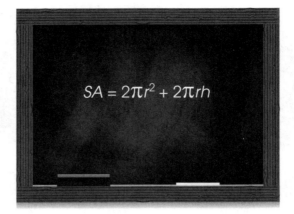

$$SA = 2\pi r^2 + 2\pi rh$$

Don't let the fancy expression scare you.

That just means $2 \times \pi \times r \times r$ plus $2 \times \pi \times r \times h$!

Examples

Now that we've got the formula, let's put it to work.

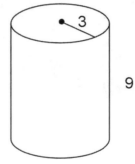

This cylinder has a radius of 3 and a height of 9. Use the surface area formula:

$$\text{Surface area} = (2 \times \text{base area}) + (\text{circumference} \times \text{height})$$
$$= (2 \times \pi \times r \times r) + (2 \times \pi \times r \times h)$$
$$= (2 \times \pi \times 3 \times 3) + (2 \times \pi \times 3 \times 9)$$
$$= (2 \times \pi \times 9) + (2 \times \pi \times 27)$$
$$= (18\pi) + (54\pi)$$
$$= 72\pi$$

The surface area is 72π square units.

egghead's Guide to Geometry

Chapter 12: Cylinders

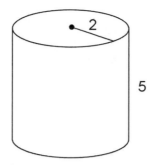

This cylinder has a radius of 2 and a height of 5. Use the surface area formula:

$$\text{Surface area} = (2 \times \text{base area}) + (\text{circumference} \times \text{height})$$
$$= (2 \times \pi \times r \times r) + (2 \times \pi \times r \times h)$$
$$= (2 \times \pi \times 2 \times 2) + (2 \times \pi \times 2 \times 5)$$
$$= (2 \times \pi \times 4) + (2 \times \pi \times 10)$$
$$= (8\pi) + (20\pi)$$
$$= 28\pi$$

The surface area is 28π square units.

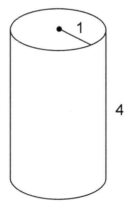

This cylinder has a radius of 1 and a height of 4. Use the surface area formula:

$$\text{Surface area} = (2 \times \text{base area}) + (\text{circumference} \times \text{height})$$
$$= (2 \times \pi \times r \times r) + (2 \times \pi \times r \times h)$$
$$= (2 \times \pi \times 1 \times 1) + (2 \times \pi \times 1 \times 4)$$
$$= (2 \times \pi \times 1) + (2 \times \pi \times 4)$$
$$= (2\pi) + (8\pi)$$
$$= 10\pi$$

The surface area is 10π square units.

Practice Questions

1. What is the surface area of the cylinder shown?

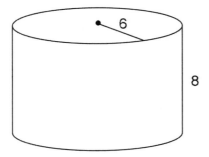

2. What is the surface area of the cylinder shown?

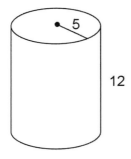

3. What is the surface area of a cylinder with a radius of 4 inches and a height of 7 inches?

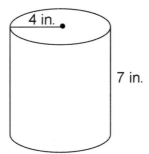

4. The cylinder shown has a radius of 3.2 millimeters and a height of 11 millimeters. Find its surface area.

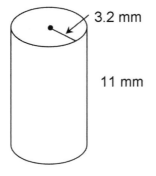

5. The cylinder shown has a radius of 2 feet and a height of 13 feet. What is its surface area?

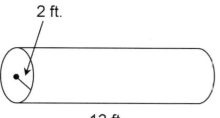

2 ft.

13 ft.

 **Solutions**

1. This cylinder has a radius of 6 and a height of 8. Use the surface area formula:

$$\text{Surface area} = (2 \times \text{base area}) + (\text{circumference} \times \text{height})$$
$$= (2 \times \pi \times r \times r) + (2 \times \pi \times r \times h)$$
$$= (2 \times \pi \times 6 \times 6) + (2 \times \pi \times 6 \times 8)$$
$$= (2 \times \pi \times 36) + (2 \times \pi \times 48)$$
$$= (72\pi) + (96\pi)$$
$$= 168\pi$$

The surface area is 168π square units.

2. This cylinder has a radius of 5 and a height of 12. Use the surface area formula:

$$\text{Surface area} = (2 \times \text{base area}) + (\text{circumference} \times \text{height})$$
$$= (2 \times \pi \times r \times r) + (2 \times \pi \times r \times h)$$
$$= (2 \times \pi \times 5 \times 5) + (2 \times \pi \times 5 \times 12)$$
$$= (2 \times \pi \times 25) + (2 \times \pi \times 60)$$
$$= (50\pi) + (120\pi)$$
$$= 170\pi$$

The surface area is 170π square units.

Solids

3. The surface area of the cylinder is 88π inches^2.

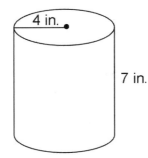

The radius of the cylinder is 4 inches, and its height is 7 inches. Plug these into the formula:

$$\text{Surface area} = (2 \times \text{base area}) + (\text{circumference} \times \text{height})$$
$$= (2 \times \pi \times r \times r) + (2 \times \pi \times r \times h)$$
$$= (2 \times \pi \times 4 \times 4) + (2 \times \pi \times 4 \times 7)$$
$$= (2 \times \pi \times 16) + (2 \times \pi \times 28)$$
$$= (32\pi) + (56\pi)$$
$$= 88\pi$$

4. The surface area of the cylinder is 90.88π square millimeters.

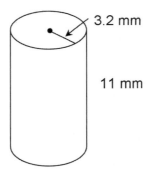

$$\text{Surface area} = (2 \times \text{base area}) + (\text{circumference} \times \text{height})$$
$$= (2 \times \pi \times r \times r) + (2 \times \pi \times r \times h)$$
$$= (2 \times \pi \times 3.2 \times 3.2) + (2 \times \pi \times 3.2 \times 11)$$
$$= (2 \times \pi \times 10.24) + (2 \times \pi \times 35.20)$$
$$= (20.48\pi) + (70.40\pi)$$
$$= 90.88\pi$$

5. The surface area of the cylinder is 60π square feet.

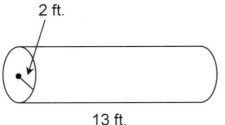

$$\text{Surface area} = (2 \times \text{base area}) + (\text{circumference} \times \text{height})$$
$$= (2 \times \pi \times r \times r) + (2 \times \pi \times r \times h)$$
$$= (2 \times \pi \times 2 \times 2) + (2 \times \pi \times 2 \times 13)$$
$$= (2 \times \pi \times 4) + (2 \times \pi \times 26)$$
$$= (8\pi) + (52\pi)$$
$$= 60\pi$$

egghead's Guide to Geometry

Chapter Review

1. Find the volume of the cylinder shown.

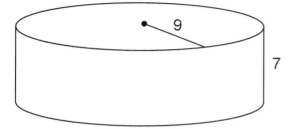

2. What is the volume of the cylinder shown?

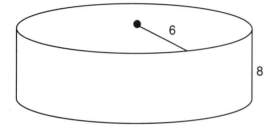

3. What is the volume of the cylinder shown?

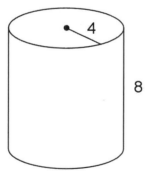

4. Find the surface area of the cylinder shown.

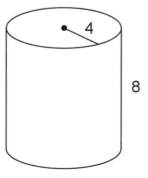

Solids

5. Find the volume and surface area of the cylinder shown.

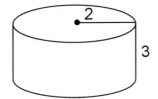

6. What is the volume of the cylinder shown?

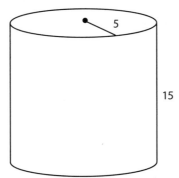

7. A grain silo has a radius of 3 meters, as shown. If the silo holds exactly 54π cubic meters of grain, how tall is the silo?

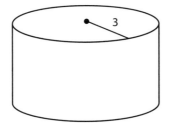

8. Find the surface area of the cylinder shown.

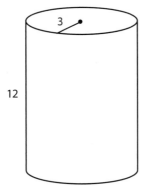

9. Bales Auto Shop stores its spare parts in a cylindrical drum that measures 10 feet from the center of the top of the drum to its outer edge, as shown. The drum has a surface area of 360π square feet. How tall is the drum?

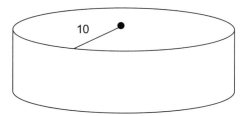

10. The cylinder shown in the figure has a diameter of 12 centimeters and a height of 18 centimeters. Find the volume and surface area of the cylinder.

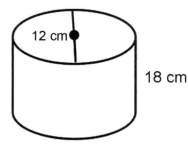

11. Joshua has a metal canister in the shape of a cylinder. Its radius is 7 inches, and its height is 21 inches. Joshua wants to wrap the canister in aluminum foil. How much foil will he need to cover the entire canister?

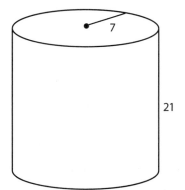

Solids

12. What is the surface area of the cylinder shown?

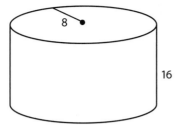

13. A water tank in the shape of a cylinder has a radius of 3.5 yards and a height of 10 yards. How much water, in cubic yards, can the water tank hold?

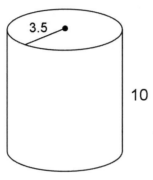

14. A coffee mug measures 10 centimeters in diameter from inside edge to inside edge. If the mug holds 275π cubic centimeters of coffee, what is the height of the mug?

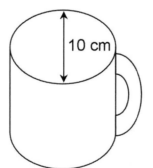

15. A concrete pipe measures 20 feet in length. The pipe is filled exactly halfway with water, as shown by the dotted lines. If the volume of water in the pipe is 160π cubic feet, find the pipe's diameter.

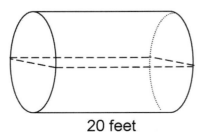

20 feet

Chapter 12: Cylinders

Solutions

1. To find the volume, we multiply $\pi \times r \times r \times h$. For this cylinder, the radius is 9 and the height is 7.

$$\text{Volume} = \text{area of base} \times \text{height}$$
$$= \pi \times r \times r \times h$$
$$= \pi \times 9 \times 9 \times 7$$
$$= 567\pi$$

The volume is 567π cubic units.

2. For this cylinder, the radius is 6 and the height is 8. Multiply $\pi \times r \times r \times h$:

$$\text{Volume} = \text{area of base} \times \text{height}$$
$$= \pi \times r \times r \times h$$
$$= \pi \times 6 \times 6 \times 8$$
$$= 288\pi$$

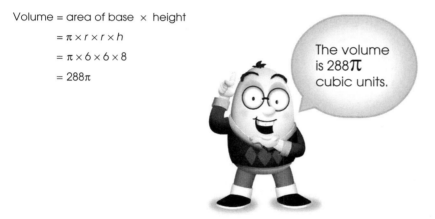

The volume is 288π cubic units.

3. To find the volume, we multiply $\pi \times r \times r \times h$. For this cylinder, the radius is 4 and the height is 8.

$$\text{Volume} = \text{area of base} \times \text{height}$$
$$= \pi \times r \times r \times h$$
$$= \pi \times 4 \times 4 \times 8$$
$$= 128\pi$$

The volume is 128π cubic units.

4. This cylinder has a radius of 4 and a height of 8. Use the surface area formula:

$$\text{Surface area} = (2 \times \text{base area}) + (\text{circumference} \times \text{height})$$
$$= (2 \times \pi \times r \times r) + (2 \times \pi \times r \times h)$$
$$= (2 \times \pi \times 4 \times 4) + (2 \times \pi \times 4 \times 8)$$
$$= (2 \times \pi \times 16) + (2 \times \pi \times 32)$$
$$= (32\pi) + (64\pi)$$
$$= 96\pi$$

The surface area is 96π square units.

Solids

5.

$$V = \pi \times r^2 \times h \qquad\qquad SA = (2 \times \pi \times r \times r) + (2 \times \pi \times r \times h)$$

For this cylinder, the radius is 2 and the height is 3. To find the volume, multiply $\pi \times r \times r \times h$:

$$
\begin{aligned}
\text{Volume} &= \text{area of base} \times \text{height} \\
&= \pi \times r \times r \times h \\
&= \pi \times 2 \times 2 \times 3 \\
&= \pi \times 4 \times 3 \\
&= 12\pi
\end{aligned}
$$

The volume is 12π cubic units.

To find the surface area, use the surface area formula:

$$
\begin{aligned}
\text{Surface area} &= (2 \times \text{base area}) + (\text{circumference} \times \text{height}) \\
&= (2 \times \pi \times r \times r) + (2 \times \pi \times r \times h) \\
&= (2 \times \pi \times 2 \times 2) + (2 \times \pi \times 2 \times 3) \\
&= (2 \times \pi \times 4) + (2 \times \pi \times 6) \\
&= (8\pi) + (12\pi) \\
&= 20\pi
\end{aligned}
$$

The surface area is 20π square units.

6. The volume is 375π.

The radius of this cylinder is 5 and the height is 15. Multiply $\pi r^2 h$:

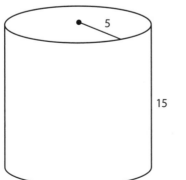

$$
\begin{aligned}
\text{Volume} &= \text{area of base} \times \text{height} \\
&= \pi \times r \times r \times h \\
&= \pi \times 5 \times 5 \times 15 \\
&= 375\pi
\end{aligned}
$$

egghead's Guide to Geometry

Chapter 12: Cylinders

7. The height of the silo is 6 meters.

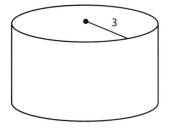

In this question, we are given the radius and volume of the silo and asked to find its height. Start with the volume formula and plug in the values given:

$$\text{Volume} = \text{area of base} \times \text{height}$$
$$54\pi = \pi \times 3 \times 3 \times h$$
$$54\pi = \pi \times 9 \times h$$
$$\frac{54\pi}{9\pi} = h$$
$$6 = h$$

8. The surface area is 90π square units.

This cylinder has a radius of 3 and a height of 12. Use the surface area formula:

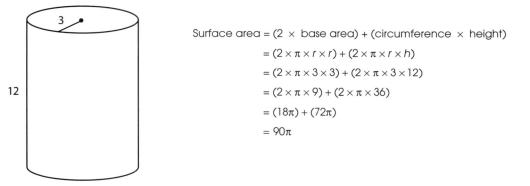

$$\text{Surface area} = (2 \times \text{base area}) + (\text{circumference} \times \text{height})$$
$$= (2 \times \pi \times r \times r) + (2 \times \pi \times r \times h)$$
$$= (2 \times \pi \times 3 \times 3) + (2 \times \pi \times 3 \times 12)$$
$$= (2 \times \pi \times 9) + (2 \times \pi \times 36)$$
$$= (18\pi) + (72\pi)$$
$$= 90\pi$$

9. The drum is 8 feet tall.

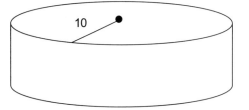

According to the question, the drum measures 10 feet from the center of its top to its outer edge. Therefore, the drum has a radius of 10 feet. Its surface area is 360π square feet. Using the surface area formula, plug in the given values and work backward to find the height:

$$\text{Surface area} = (2 \times \text{base area}) + (\text{circumference} \times \text{height})$$
$$360\pi = (2 \times \pi \times r \times r) + (2 \times \pi \times r \times h)$$
$$360\pi = (2 \times \pi \times 10 \times 10) + (2 \times \pi \times 10 \times h)$$
$$360\pi = (2 \times \pi \times 100) + (20 \times \pi \times h)$$
$$360\pi = (200\pi) + (20\pi h)$$
$$360\pi - 200\pi = 20\pi h$$
$$160\pi = 20\pi h$$
$$\frac{160\pi}{20\pi} = h$$
$$h = 8$$

The height measures 8 feet.

Solids

10. The cylinder has a diameter of 12 centimeters. So, its radius measures 6 centimeters. To find the volume and surface area, apply the formulas as shown:

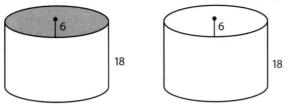

This cylinder has a height of 18 centimeters. Its volume is 648π cubic centimeters:

$$\text{Volume} = \text{area of base} \times \text{height}$$
$$= \pi \times r \times r \times h$$
$$= \pi \times 6 \times 6 \times 18$$
$$= 648\pi$$

Its surface area is 288π square centimeters:

$$\text{Surface area} = (2 \times \text{base area}) + (\text{circumference} \times \text{height})$$
$$= (2 \times \pi \times r \times r) + (2 \times \pi \times r \times h)$$
$$= (2 \times \pi \times 6 \times 6) + (2 \times \pi \times 6 \times 18)$$
$$= (2 \times \pi \times 36) + (2 \times \pi \times 108)$$
$$= (72\pi) + (216\pi)$$
$$= 288\pi$$

11. Joshua will need 392π square inches of foil to cover the canister.

This cylinder has a radius of 7 inches and a height of 21 inches. Use the surface area formula:

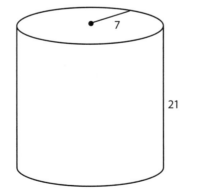

$$\text{Surface area} = (2 \times \text{base area}) + (\text{circumference} \times \text{height})$$
$$= (2 \times \pi \times r \times r) + (2 \times \pi \times r \times h)$$
$$= (2 \times \pi \times 7 \times 7) + (2 \times \pi \times 7 \times 21)$$
$$= (2 \times \pi \times 49) + (2 \times \pi \times 147)$$
$$= (98\pi) + (294\pi)$$
$$= 392\pi$$

The surface area of the canister is 392π in^2.

12. The surface area is 384π.

This cylinder has a radius of 8 and a height of 16. Use the surface area formula:

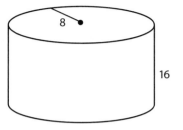

Surface area = $(2 \times \text{base area}) + (\text{circumference} \times \text{height})$

$= (2 \times \pi \times r \times r) + (2 \times \pi \times r \times h)$

$= (2 \times \pi \times 8 \times 8) + (2 \times \pi \times 8 \times 16)$

$= (2 \times \pi \times 64) + (2 \times \pi \times 128)$

$= (128\pi) + (256\pi)$

$= 384\pi$

13. The tank can hold 122.5π cubic yards of water.

To answer this question, find the volume of the tank. The tank has a radius of 3.5 yards and a height of 10 yards. Multiply $\pi \times r \times r \times h$:

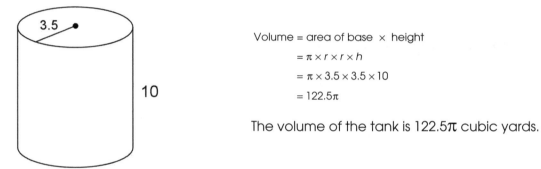

Volume = area of base $\times$ height

$= \pi \times r \times r \times h$

$= \pi \times 3.5 \times 3.5 \times 10$

$= 122.5\pi$

The volume of the tank is 122.5π cubic yards.

14. The height of the mug is 11 centimeters.

The volume of the mug is 275π centimeters, and its diameter is 10 centimeters. This means its radius is 5 centimeters. Plug these values into the volume formula and solve for the height:

Volume = area of base $\times$ height

$275\pi = \pi \times r \times r \times h$

$275\pi = \pi \times 5 \times 5 \times h$

$275\pi = 25\pi \times h$

$h = \dfrac{275\pi}{25\pi}$

$h = 11$

Solids

15. The diameter of the pipe is 8 feet.

When the pipe is half full, it holds 160π ft.^{3} of water. To find the total volume of the pipe, double this figure: the volume is 320π ft^3.

We are told that the pipe measures 20 feet in length. Using the volume formula, solve for the radius of the pipe:

$$\text{Volume} = \text{area of base} \times \text{height}$$
$$320\pi = \pi \times r \times r \times 20$$
$$320\pi = \pi \times r^2 \times 20$$
$$320\pi = 20\pi \times r^2$$
$$\frac{320\pi}{20\pi} = r^2$$
$$16 = r^2$$
$$\sqrt{16} = \sqrt{r^2}$$
$$4 = r$$

If the radius of the pipe is 4 feet, the diameter of the pipe must be 8 feet.

NOTES

NOTES

NOTES

NOTES

NOTES

NOTES

NOTES